Intellectual Property Rights Management in Developing Countries

About the Centre

The Centre for Science and Technology of the Non-Aligned and Other Developing Countries (NAM S&T Centre) is an inter-governmental organisation with a membership of 45 countries spread over Asia, Africa, Middle East and Latin America. Besides this, 12 S&T agencies and academic/research institutions of Bolivia, Botswana, Brazil, India, Nigeria and Turkey are the members of the S&T-Industry Network of the Centre. The Centre was set up in 1989 to promote South-South cooperation through mutually beneficial partnerships among scientists and technologists and scientific organisations in developing countries. It implements a variety of programmes including international workshops, meetings, roundtables, training courses and collaborative projects and brings out scientific publications, including a quarterly Newsletter. It is also implementing 4 Fellowship schemes, namely, NAM S&T Centre Research Fellowship, Joint NAM S&T Centre – ICCBS Karachi Fellowship, Joint CSIR/CFTRI (Diamond Jubilee) - NAM S&T Centre Fellowship and Joint NAM S&T Centre – ZMT Bremen Fellowship. These activities provide, among others, the opportunity for scientist-to-scientist contact and interaction, training and expert assistance, familiarising the scientific community on the latest developments and techniques in the specific subject areas, and identification of technologies for transfer between member countries. The Centre has so far brought out 49 publications and has organised 80 international workshops and training programmes.

For further details, please visit www.namstct.org or write to the Director, NAM S&T Centre, Core 6A, 2nd Floor, India Habitat Centre, Lodhi Road, New Delhi-110003, India (Phone: +91-11-24645134/24644974; Fax: +91-11-24644973; E-mail: namstct@gmail.com; namstct@bol.net.in).

Intellectual Property Rights Management in Developing Countries

— *Editors* —
Sheila Mavis Nyatlo
Catherina Caballero
M.A.M. Fernando

CENTRE FOR SCIENCE & TECHNOLOGY OF THE
NON-ALIGNED AND OTHER DEVELOPING COUNTRIES
(NAM S&T CENTRE)

2012
DAYA PUBLISHING HOUSE®
New Delhi - 110 002

ISBN 9789351240846

Centre for Science and Technology of the Non-Aligned and Other Developing Countries (NAM S&T Centre)
Core-6A, 2nd Floor, India Habitat Centre, Lodhi Road,
New Delhi-110 003 (India)
Phone: +91-11-24644974, 24645134, Fax: +91-11-24644973
E-mail: namstct@gmail.com
Website: www.namstct.org

Published by : **Daya Publishing House®**
A Division of
Astral International Pvt. Ltd.
– ISO 9001:2008 Certified Company –
4760-61/23, Ansari Road, Darya Ganj,
New Delhi - 110 002
Phone: 23245578, 23244987
Fax: (011) 23260116
e-mail : dayabooks@vsnl.com website :
www.dayabooks.com

Laser Typesetting : **Classic Computer Services**
Delhi - 110 035

Printed at : **Chawla Offset Printers**
Delhi - 110 052

PRINTED IN INDIA

Foreword

In the current competitive and knowledge driven global economic environment, intellectual property rights, which are the results of human creative efforts, have become unprecedentedly relevant and important to any country, whether industrialized or developing. The future of the developing world would largely depend on their ability to create and generate knowledge and the resultant intellectual property rights thereby narrowing the knowledge gap and enhancing the use of knowledge for economic growth.

As the subject of intellectual property spreads across many facets of human life such as innovation and creativity, trade, industry, environment, health, access to knowledge and food security, it is now essential to make the Intellectual Property system, in any given country, community associated and development oriented. The intellectual property rights, if duly managed and used, can extensively contribute to the economic growth and poverty alleviation. Thus, the promotion of innovation and creativity in the developing countries, particularly aiming the knowledge related and technology intensive industries demands high priority. Moreover, the effective protection and management of national Intellectual Property assets such as geographical indications, traditional knowledge, traditional creative expressions and genetic resources require constructive attention.

Having realized the importance of the role of Intellectual Property in socio-economic advancement, most of the developing countries are in the process of building up or updating their Intellectual Property systems in line with domestic needs and international standards. They are making attempts to respond to the new order and address the Intellectual Property issues - both national and global - with multi-faceted approaches as well as expectations. This process involves many challenges. In meeting those challenges, the developing economies must not only to be innovative and forward looking but also be mindful of the increased south – south cooperation. The concerted efforts will facilitate the establishment of strong national Intellectual Property systems helping them to achieve public policy needs and economic development goals.

In this context, the organization of the 'International Advanced Training Course on Intellectual Property Rights (IPR) for NAM and Other Developing Countries' during July 12-17, 2010 in New Delhi/Manesar, India by the Centre for Science and Technology of the Non-aligned and Other Developing Countries (NAM S&T Centre) in association with the Patent Facilitating Centre (PFC) of the Technology Information, Forecasting and Assessment Council (TIFAC), Government of India, was timely and futuristic. Based on the deliberations at the training course, the NAM S&T Centre has brought out this book containing 21 papers authored by the intellectuals and experts mainly from the developing countries for wider dissemination of knowledge and experience on Intellectual Property. It is also a highly welcome step forward for enhanced South–South cooperation. The book will be useful to those interested not only in Intellectual Property rights but also in development oriented activities in general. I sincerely congratulate the authors, editors, publisher and others responsible for its publication on their noble efforts and grand success.

D.M. Karunaratna

LL.B. (Hons), M.Phil, Ph.D.
Attorney at Law
Director General of Intellectual Property
Intellectual Property Office of Sri Lanka

From the Editors' Desk

The relationship between Intellectual Property Rights (IPRs) and development is very complex indeed. On one hand, there are theoretical arguments suggesting that stronger IPRs can have positive effects on development. On the other hand, there are theoretical arguments against stronger IPRs in developing countries.

However, amidst all of the arguments; developing countries are trading with developed countries that utilise IPR as a source of competitive advantage; this then poses a challenge for developing countries and urges them to recognise Intellectual Property (IP) as the tool for competitive advantage.

The challenge for developing countries is converting what comes out of basic research into tangible outcomes, to which real value has been added, or beneficiating indigenous knowledge and packaging it in a way that would generate value.

Majority of developing countries have widespread genetic resources from plants, animals and micro-organisms, amounting to 90 per cent of the world's genetic resources (GR). Communities and individuals in developing countries have exploited these GR through generations. Their use is embodied in what often is referred to as Traditional Knowledge (TK). However, the use of such knowledge and resources is not limited to local contexts, and many innovations relate to and draw on them. The broader use of TK and GR raises the prospect that they may play an important role in driving growth in developing countries. Therefore, the main issue is how this prospect might be best realised, particularly when the exploitation of TK and GR is coming increasingly under the governance of various, and sometimes conflicting IPR frameworks.

It is however encouraging to note that developing countries are working towards ensuring that their IPR are protected and are implementing mechanisms to assist with the exploitation thereof in realising the return on investment for their countries.

Critical shortage of skills in terms of trained individuals having practical understanding of IP in relation to the development of knowledge based economies and personnel who know how to assess the value of IP is always a challenge facing developing countries.

Governments in developing countries are frequently encouraged to focus upon educating people to make IP and its role in economic development more understandable especially in terms of research institutions and provision of insights into the use of the IP system.

Developing countries should then focus on identifying and developing their strengths, projecting them to be a sustainable source of competitive advantage.

Sheila Mavis Nyatlo

Catherina Caballero

M.A.M. Fernando

Introduction

Intellectual Property Right (IPR) is an exclusive right of wealth created and legally possessed by people engaged in the field of science, technology, literature and art. In today's globalised competitive society, an IPR system focusing on patent, trademark, copyrights, etc. is playing an increasingly important role in economic and social development. In this connection, the developing countries in particular have several challenges to overcome while evolving their own IP systems in order to satisfy their public policy needs and socio-economic developmental goals. Over the years, these countries have responded to the new order in different ways and have tried to handle global IPR issues with a sense of commonalty; especially in the matters related to protection of genetic resources and traditional knowledge. There is a sense of urgency in developing IP systems in each country to derive maximum advantage from the IPR system to address efficient utilisation of innovations and creativity.

Looking into the increasing significance of IPR related issues and assessing the need of direct participation of scientific community of the developing world in legal management of the knowledge of science and innovations, the Centre for Science and Technology of the Non-aligned and Other Developing Countries (NAM S&T Centre) in association with the Patent Facilitating Centre (PFC) of the Technology Information, Forecasting and Assessment Council (TIFAC), Government of India organised a six days 'International Advanced Training Course on Intellectual Property Rights (IPR) for NAM and Other Developing Countries' during 12-17 July 2010 in New Delhi/ Manesar, India. The theme of the training course was 'Development of IPR Systems for National Growth' and its objective was to create awareness among scientific community about the legal framework of scientific inventions and practices in developing countries. The programme was supported by the Department of Science and Technology (DST), Ministry of Science and Technology, Government of India.

The training course was attended by 49 trainee participants from 28 countries, including Botswana, Brazil, Cambodia, Colombia, Egypt, Ghana, India, Indonesia, Iran, Iraq, Jamaica, Kenya, Malawi, Malaysia, Mexico, Myanmar, Nepal, Nigeria,

Pakistan, Panama, Peru, South Africa, Sri Lanka, Sudan, Tanzania, Turkey, Uganda and Zambia, of which 18 delegates were from the host country India and the remaining 31 represented other developing countries. There were 11 resource persons/trainers from India, who delivered 17 lectures on various themes related to IPR. The overseas trainee participants were from Botswana [Mr. Botshelo Maedza, Acting Principal Civil Engineer, Botswana Technology Centre (BOTEC), Gaborone and Mr. Letsomo la Tshipa, Ass. Performing Arts Officer (AAW – Copyright), Department of Arts and Culture, Ministry of Youth, Sports and Culture, Gaborone]; Brazil [Dr. Marylin Peixoto da Silva Nogueira, Director, ICT Policies and Sectoral Programmes, Ministry of Science and Technology, Brasilia]; Cambodia [Mr. Chhean Piseth, Chief Officer, Department of Industrial Property, Ministry of Industry, Mines and Energy, Phnom Penh]; Colombia [Ms. Maria Consuelo Velásquez, Intellectual Property Advisor, Administrative Department of Science, Technology and Innovation, Columbian Institute for the Development of Science and Technology (COLCIENCIAS), Bogotá]; Egypt [Ms. Handra Mohamed Mohamed Nagy, Access To Knowledge Project Officer, Alexandria]; Ghana [Ms. Nana Ama Akyiaa Prempeh, Principal State Attorney, Registrar General's Department, Ministry of Justice, Kumasi]; Indonesia [Dr. Anny Sulaswatty, Head of Legal and Public Relation Affair, Ministry of Research and Technology, Jakarta]; Iran [Dr.. M. Reza Bakhtiari, Director and Head of Intellectual Property Office, Iranian Research Organization for Science and Technology (IROST), Tehran]; Iraq [Miss Teriza Jasim Ridha, Director of IPO, Central Organization for Standardization and Quality Control (COSQC), Baghdad]; Jamaica [Mrs. Yvette Sutherland-Reid, Senior Legal Officer/Attorney-at-law, Ministry of Industry, Investment and Commerce, Kingston]; Kenya [Mr. Kioko Nzuki Mwania, Principal Research Officer, Directorate of Research Management and Development (DRMD), Ministry of Higher Education, Science and Technology, Nairobi]; Malawi [Mr. Kudonda Fexter Nseula, Technology Transfer Officer (Development) and Mr. Petros Orton Shiyombo Musukwa, Technology Transfer Officer (Dissemination), National Commission for Science and Technology (NCST), Lilongwe]; Malaysia [Mr. Ahmad Fareed bin Abdul Rahman, Assistant Secretary, Ministry of Science, Technology and Innovation, Putrajaya]; Mexico [Ms. Ana Paula Ramírez Ramírez, International Affairs Coordinator, Tequila Regulatory Council, Jalisco]; Myanmar [Dr. Phyo Mon Win, Assistant Director, Intellectual Property Section, Ministry of Science and Technology, Nay Pyi Taw]; Nepal [Dr. Chiranjivi Regmi, Chief, Faculty of Technology, Nepal Academy of Science and Technology (NAST), Kathmandu]; Nigeria [Dr. Martins Ochubiojo Emeje, Research Fellow and IP and Technology Transfer Coordinator, Biophysical Drug Delivery Unit, National Institute for Pharmaceutical Research and Development (NIPRD), Abuja]; Pakistan [Ms. Syeda Rumla Naqvi, Industrial Liaison Officer, Pakistan Council of Scientific and Industrial Research (PCSIR) Labs, Karachi]; Panama [Dr. Catherina Caballero-George, Staff Scientist, Unit of Molecular Pharmacology and Pharmacognosy, Institute for Scientific Research and High Technology Services (INDICASAT), Clayton]; Peru [Mr. Carlos A. Irigoyen, Deputy Chief of Mission, Embassy of Peru in New Delhi]; South Africa [Mrs. Sheila Mavis Nyatlo, Deputy Director, Department of Science and Technology, Pretoria]; Sri Lanka [Mr. Merengnege Antony Mervyn Fernando, Senior Mechanical Engineer, National Engineering Research and Development Centre (NERDC), Ekala, Ja Ela]; Sudan [Mr.

Ghazafi Abdalla Mohamed Ali, Counsellor, Sudanese Ministry of Foreign Affairs, Khartoum]; Tanzania [Ms. Judith Francis Kadege, Research Officer (IP Law), Tanzania Commission for Science and Technology, Dar-Es-Salaam]; Turkey [Dr. Burcun Dikmen, Chief Researcher, Scientific and Technical Research Council of Turkey (TÛBÌTAK) and Ms. Özge Eþan, Attorney at Law, Patent and Trademark Attorney, Legal Department/IPR Office, TÛBÌTAK, Ankara]; Uganda [Mr. Stephen Rwagweri, Executive Director, Engubu Za Tooro, Fort Portal and Mr. Frederick Okumu, Network Officer, Uganda Industrial Research Institute (UIRI), Kampala]; and Zambia [Mr. Lloyd Samson Thole, Assistant Registrar (IP), Patents and Companies Registration Office, Lusaka].

The Indian trainee participants were Ms. Anshul Fuloria, Ms. Aruna, Ms. Dipti, Dr. Deepika Singh, Mrs. Deepika Chaturvedi, Mrs. Geeta, Ms. Shikha Tejswi, Ms. Shashank, Ms. Shivani Shrivastava, Dr. Rollie Verma, Ms. Raunak Dubey, Mrs. Neeti Sahu and Ms. Divya Pandey from PFC-TIFAC; Ms. Debanjana Dey from the National Institute of Science, Technology and Development Studies (NISTADS), New Delhi; and Mr. Gaurav Gaur, Ms. Chitra Prasad, Ms. Priyanka Rana and Ms. Isha Parmar, Research Assistants from NAM S&T Centre.

During the Training course, lectures were delivered by eminent Indian experts and professionals Mr. R. Saha, Director, PFC-TIFAC, New Delhi [on 'Basics of IPR Management', 'International Treaties', 'IPR Audit', 'IPR in International R&D Collaborations', 'IPR Policy Development in Institutions', 'TRIPS and Emerging Issues, and 'Valuation of IPR']; Mr. A. K. Garg, Director, Department of Information Technology (DIT), Government of India [on 'Protecting Software Inventions']; Prof. V.C. Vivekanandan, NALSAR University of Law, Hyderabad [on 'International Disputes in IPR-DSB Approach']; Mr. S. Majumdar, Patent Attorney, S. Majumdar and Co., Kolkata [on 'IPR Portfolio Management']; Mr. Rajendra Kumar, Attorney, K&S Partners, Gurgaon, India [on 'Protection and Management of Geographical Indications']; Dr. S. Mauria, ADG (IPR and Policy), Indian Council of Agricultural Research (ICAR) [on 'Practices in Protection of New Plant Varieties']; Mr. V. K. Gupta, Director TKDL, Council of Scientific and Industrial Research (CSIR), New Delhi, India [on 'IPR and Traditional Knowledge']; Dr. K. Satyanarayana, Sr. DDG, Indian Council of Medical Research (ICMR), Government of India [on 'IPR Issues in Medicines – Case of Drugs Protection from Developing Countries']; Mr. Yashawant D. Panwar, Scientist-D, PFC-TIFAC, New Delhi [on 'Patentability Analysis including Patent Search']; Prof. Madhukar Sinha, WTO Study Centre, Indian Institute of Foreign Trade, New Delhi [on 'Digital Rights Management']; and Mr. V. Lakshmikumaran, Sr. Attorney, Lakshmikumaran and Sridharan Associates, New Delhi [on 'Principles and Practices of IPR Licensing'].

The overseas participants presented their country/institutional status reports in two sessions respectively chaired by Dr. A. K. Sood, Head, International Division and Adviser, DST, Government of India and Dr. Balram Sharma, Indian Agricultural Research Institute (IARI), New Delhi.

Field visits were undertaken by the participants to PFC-TIFAC office at Vishwakarama Bhawan, New Delhi; Intellectual Property Office at Dwarka, New

Delhi and other places. During this visit, Mr. R. Saha, Director, introduced the participants with the functions of the PFC and its increasing popularity in the recent years and a lecture was delivered by Ms. Sangeeta Nagar, Principal Scientific Officer on the 'Women Scientist Programme' being executed by PFC-TIFAC followed by questions and answers session. The visitors were specifically interested in knowing how they may initiate similar programmes in their own countries.

In the field visit to the Indian Patent Office (IPO), New Delhi, Dr. K.S. Kardam, Deputy Controller of Patents and Designs and Mr. S. K. Swarup, Assistant Controller of Patents and Designs presented an overview of the operations and services being executed by the IPO including its digitised information system and easy-to-access website.

At the Concluding Session Dr. Rajiv Sharma, Adviser, International Division, DST, Government of India was the Chief Guest. Mr. R. Saha, Director, PFC-TIFAC presented a summary of the Training Course followed by a Feedback Session chaired by the undersigned in which participants gave their valuable and critical comments on the course content and made suggestions for designing similar programmes in future. Some participants even mentioned that the training course organised in Manesar was very different and more beneficial to the developing countries than the ones being organised by the agencies such as WIPO and ARIPO.

The brainstorming discussion among the participants and the trainers was finally concluded with significant outcome, and to ensure the sustainability the participants, trainers and organisers unanimously adopted the 'Manesar Resolutions',

I acknowledge with gratitude the deep involvement and determined efforts of Mrs. Sheila Mavis Nyatlo, Deputy Director, Department of Science and Technology, Pretoria, South Africa, Dr. Catherina Caballero, Staff Scientist, Unit of Molecular Pharmacology and Pharmacognosy, Institute for Scientific Research and Technology Services (INDICASAT), Clayton, Panama and Mr. Merengnege Antony Mervyn Fernando, Senior Mechanical Engineer, National Engineering Research and Development Centre, Ekala, Ja Ela, Sri Lanka in editing this valuable publication. I am also indebted to the entire team of the NAM S&T Centre, particularly Mr. M. Bandyopadhyay, Dr. V.P. Kharbanda, Mr. Gaurav Gaur, Mr. Yasir Abbas Rizvi, Ms. Chitra Prasad, Ms. Priyanka Rana, Ms. Isha Parmar and Mr. Pankaj Buttan in compiling the presented papers, liaising with the authors and editors and giving a shape to this volume.

I trust that the material presented in this volume will be highly useful not only for the scientists and professionals engaged in the field of IPR, but also for the concerned authorities of the developing countries in taking appropriate policy measures to create awareness about these rights.

Prof. Arun P. Kulshreshtha

Director and Executive Head,
NAM S&T Centre

Contents

Foreword *v*
D.M. Karunaratna

From the Editors' Desk *vii*
Sheila Mavis Nyatlo, Catherina Caballero, M.A.M. Fernando

Introduction *ix*
Arun P. Kulshreshtha

Section I: African Continent

1. **South African Intellectual Property Rights Status Report** 3
Sheila Mavis Nyatlo (South Africa)

2. **Intellectual Property (IP) in Botswana: Research but No Patents** 15
Botshelo Maedza (Botswana)

3. **Intellectual Property Rights: Situation in Botswana** 21
Letsomo la Tshipa (Botswana)

4. **Country Report on IP Policy and IP Related Laws** 23
Petros Orton Shiyombo Musukwa (Malawi)

5. **Zambia: Status Paper on Intellectual Property** 26
Thompson Namukaba (Zambia)

6. **Intellectual Property Management in Zambia** 32
Lloyd Samson Thole (Zambia)

7. **IPR Management in Tanzania: The Role of Tanzania Intellectual Property Advisory Services and Information Centre (TIPASIC)** 43
Judith F. Kadege (Tanzania)

8. **Protection of Traditional Knowledge in Uganda** 57
Stephen Rwagweri Atwooki (Uganda)

9. **Intellectual Property** 65
Frederick Okumu (Uganda)

10. **The Scientists and IPR in a Resource Starved Nation: NIPRD IPTTO Experience** 70
Emeje, Martins and Gamaniel Karnyius (Nigeria)

Section II: American Continent

11. **The Intellectual Property Rights in Mexico: The Tequila Case** 79
Ana Paula Ramírez Ramírez (Mexico)

12. **Intellectual Property System in Panama: Biodiversity and Traditional Knowledge** 111
Catherina Caballero-George and Leonardo Uribe (Panama)

13. **Research and Development Networks in Colombia: An Answer for the Research Ownership Exploitation Challenge** 119
Maria Consuelo Velasquez (Colombia)

Section III: Asian Continent

14. **Intellectual Property Rights in Iraq** 137
Teriza Jasim Ridha (Iraq)

15. **Intellectual Property Rights in Pakistan: Role of PCSIR for Strengthening the IPR System for National Growth** 144
Syeda Rumla Naqvi, Dr. Nighat Afza and Dr. Tanzil Haider Usmani (Pakistan)

16. **Status of Intellectual Property Rights (IPR) with Reference to Genetic Resources in Nepal** 158
Chiranjivi Regmi (Nepal)

17. **The Current Situation of Intellectual Property in the Kingdom of Cambodia** 164
Chhean Piseth (Cambodia)

18. **Lesser Known Horticultural Products as Geographical Indications in India** 167
S.K. Soam and Mujarrab Hussain (India)

19. **Ecosystem for IPR in India Especially for Patents** 178
Yashawant Dev Panwar (India)

20. **Status of the Intellectual Property System in Myanmar** 195
Phyo Mon Win (Myanmar)

21. **The Intellectual Property Regime in Sri Lanka and its New Trends** 215
Merengnege Antony Mervyn Fernando (Sri Lanka)

22. **The Status of Indonesian Intellectual Property Rights** 229
Anny Sulaswatty (Indonesia)

Section I
African Continent

Chapter 1
South African Intellectual Property Rights Status Report

Sheila Mavis Nyatlo
Department of Science and Technology, South Africa
E-mail: mavis.nyatlo@dst.gov.za

ABSTRACT

The South African patents system is governed by the Patents Act, 1978 which affords protection to patents for inventions (including patents of addition). The Companies and Intellectual Property Registration Office (CIPRO) of the Department of Trade and Industry administers South Africa's Intellectual Property Rights (IPR) Regime whilst the Department of Science and Technology is the custodian of the Intellectual Property Rights Act from Publicly Financed Research and Development, Act No. 51 of 2008.

This Act seeks to ensure that intellectual property emanating from publicly financed research and development is identified, protected, utilised and commercialised for the benefit of the people of the Republic, whether it be for a social, economic, military or any other benefit.

Various policy initiatives largely driven by South Africa's Department of Science and Technology and its instruments, have played and will continue to play an important role in the institutions contribution to South Africa's economic growth. Intellectual Property Rights, in terms of patenting activity is dependent on research activity and research output.

Keywords: *Intellectual Property, South Africa, IPR Act, NIPMO, Patents.*

Introduction

The innovation towards a knowledge-based economy plan aims to drive South Africa's transformation from a resource-based towards a knowledge-based economy;

where the production and dissemination of knowledge leads to economic benefits and enriches all fields of human endeavour. This plan, termed the 10-Year Innovation Plan (TYIP), was published by South Africa's Department of Science and Technology (DST) in 2008. The success of migration towards a knowledge-based economy is measured by the degree to which science and technology (S&T) play a leading role in enhancing productivity, economic growth and socio-economic development. The TYIP addresses an array of social, economic, political, environmental, scientific and technological benefits and is designed to stimulate multidisciplinary thinking and challenge South Africa's researchers to answer existing questions, create new disciplines and develop new technologies.

The growing commitment of resources has been accompanied by extensive policy experimentation and grounded in the overall concept of the National System of Innovation (NSI); policy design has drawn extensively on international experience and thinking (Kaplan, 2009).

South Africa's patent regime is ranked highly among developing countries in terms of strength. The courts function effectively and strongly to enforce the rights of intellectual property (IP) holders. South Africa is a signatory to most international treaties on IP Law. This includes the Paris Convention for the Protection of Industrial Property, the Patent Cooperation Treaty (PCT), the Budapest Treaty on International Recognition of the Deposit of Microorganism for the Purpose of Patent Procedure and the Trade Related Aspects of Intellectual Property Agreement. In terms of copyright and related rights, South Africa is a signatory to the Berne Convention for the Protection of Literary and Artistic Works and to the World Intellectual Property Organisation (WIPO) and the WIPO Performances and Phonograms Treaty (WPPT) (Kaplan, 2009).

Legislation

The Companies and Intellectual Property Registration Office (CIPRO), which came about after the merger between the former South African Companies Registration Office (SACRO) and the South African Patents and Trade Marks Office (SAPTO) of the Department of Trade and Industry administers South Africa's Intellectual Property Rights (IPR) regime that comprises the following legislation:

(*a*) The Patents Act, 1978 (Act No. 57 of 1978), that provides for the registration and granting of patents for inventions and for matters connected therewith. The Act was amended in 2002 and 2005.

(*b*) the Trade Marks Act, 1963 and 1993 (Act No. 62 of 1963 and Act No. 194 of 1993).

(*c*) the Designs Act, 1967 and 1993 (Act No. 57 of 1967 and Act No. 195 of 1993).

(*d*) the Copyright Act, 1978 (Act No. 98 of 1978).

(*e*) the Registration of Copyright in Cinematograph Films Act, 1977 (Act No. 62 of 1977).

(*f*) the Merchandise Marks Act, 1941 (Act No. 17 of 1941), amended 2002.

(*g*) the Intellectual Property Laws Rationalisation Act, 1977 (Act No. 107 of 1977).

CIPRO is responsible for the registration of patents, trademarks, designs and copyright. It keeps records of all patents, trademarks and designs lodged in South Africa, including details of the owners/proprietors.

Patents

The South African patents system is governed by the Patents Act, 1978 (as amended in 2002 and 2005), which affords protection to patents for inventions (including patents of addition). The absence of novelty vitiates a patent absolutely in South Africa and obvious inventions are refused protection. Unless a complete specification is filed in the first instance, a priority date is established by an application accompanied by a provisional specification. Applicants deciding to pursue protection thereafter must file a fresh application accompanied by a complete specification within 15 months of the date of filing the provisional application claiming priority from that date. South Africa operates a deposit system for patent registration with examination of compliance with administrative requirements only. No examination on the merits takes place. The patent term is 20 years from the date of filing the complete specification. A patent must be renewed every year from the fourth following the filing date. South Africa is not a member of the African Regional Intellectual Property Organization (ARIPO), an inter-governmental organisation with 16 Member States.

A provisional application is available in South Africa and a few other countries, including the United Kingdom and the United States of America, and provides provisional protection for 12 months in all Paris Convention countries (about 154). This can be followed by a complete application in selected countries via the Patent Co-operation Treaty (PCT) system (before the expiry of the 12-month period) which, if granted, provides protection generally for 20 years from the date of filing the complete application, and usually renewable from the fourth year in each country.

A South African patent has effect only in the territorial area of South Africa. If someone requires protection in other countries, they must file separate applications in the relevant countries. An exception is the regional patent for Europe and regional patents in Africa (ARIPO; African Intellectual Property Organization). The South African patent application can serve as a basis for claiming so-called convention priority in respect of foreign applications in the other convention countries.

Trademarks

The Trade Marks Act, 1993 (as amended), governs the South African trade mark registration system. Distinctive signs of all kinds and descriptions which are capable of graphical representation can be registered as trade marks. The 42 classes of goods and services of the International Classification are followed. The statute is generally compatible with the European Directive concerning trade marks and affords protection against unauthorised use of both identical and confusingly similar marks in relation to the goods or services in respect of which a mark is registered and in respect of similar or dissimilar goods or services in defined circumstances.

Designs

The Designs Act, 1993, provides protection to aesthetic and functional designs through registration. Functional designs have features which are necessitated by the function which the article to which the design is applied is to perform and include integrated circuit topography and mask works related thereto. Aesthetic designs relate to designs applied to articles of manufacture because of the pattern, shape, configuration or ornamentation of the design which is judged solely by the eye. Aesthetic designs must be new and original while functional designs must be new and not commonplace.

An aesthetic design must meet the requirements of being new and original to qualify for registration. A functional design must meet the requirements of being new and not commonplace in the art in question. In both cases, the article to which the design is applied must be intended to be multiplied by an industrial process. It is possible to obtain both aesthetic as well as functional design protection for the same design. The duration for an aesthetic design is 15 years, and 10 years for a functional design, and both periods are subject to the payment of annual renewal fees.

Copyright

Copyright is conferred by the Copyright Act, 1978, without registration. The Act protects literary, artistic and musical works, sound recordings, cinematographic films, computer programmes, radio and television broadcasts, satellite transmissions, and published editions. The categories of person who are afforded copyright protection are defined in the Act and include foreign persons and incorporated bodies from member states of the Berne Convention. Copyright comes into being automatically and no registration is required. In respect of cinematograph films, registration is possible but it is not necessary for copyright to exist.

The duration of copyright is relatively long and the term is different for different categories of works. For example:

(*a*) For literary, musical or artistic works other than photographs the term is the life of the author plus 50 years from the end of the year in which the author dies;

(*b*) For cinematograph films, photographs and computer programs, the term is 50 years from the end of the year in which the work is made available to the public or is made (where no publication takes place);

(*c*) In the case of sound recordings and published editions, the term is 50 years from the end of the year in which the recording or edition is first published.

Registration of copyright in cinematograph films is provided for by the Registration of Copyright in Cinematograph Films Act, 1977 (Act No. 62 of 1977).

Plant Breeder's Rights

The Plant Breeders' Right, 1976 (Act No. 15 of 1976) makes provision for the protection of new varieties of plant. This Act applies to varieties that have been bred by conventional means as well as varieties that have been genetically modified. To qualify for protection under the Act the variety must be new, distinct, uniform and stable.

Plant Breeders' Rights are granted for 25 years for vines and trees, and for 20 years for all other cases calculated from the date on which a certificate of registration is issued. Registration is with the Department of Forestry and Fisheries, Govt. of India.

Intellectual Property Rights from Publicly Financed Research and Development Act

The Department of Science and Technology (DST) is the custodian of the IPR from Publicly Financed Research and Development (IPR-PFRD) Act; 2008 (Act No. 51 of 2008).

The objectives of the Act are to:

(*a*) provide for more effective use of IP emanating from publicly-financed research and development (R&D);

(*b*) establish the National Intellectual Property Management Office (NIPMO) and the Intellectual Property Fund;

(*c*) provide for the establishment of offices of technology transfer at institutions; and to provide for matters connected therewith.

The Act seeks to ensure that IP emanating from publicly-financed R&D is identified, protected, used and commercialised for the benefit of the people of the Republic of South Africa, whether it be for social, economic, military or any other benefit.

The National Intellectual Property Management Office (NIPMO)

NIPMO was established in terms of the IPR-PFRD Act. NIPMO's role is to promote the objectives of the IPR-PFRD Act. This includes the statutory protection, management and commercialisation of the IP strategies around IPR to enhance protection of IP rights and ensure synergy with other policies. It will also develop national capacity to manage technology transfer and licensing and commercialisation.

Status of Intellectual Property Rights in South Africa

Table 1.1 and Figure 1.1 show the South Africa's relative position in terms of patents has declined. Its share in all foreign patents registered in the United States Patent and Trademark Office (USPTO) has consistently declined from 0,28 per cent in 1992 to 0,11 per cent in 2007. South Africa's share in PCT patent fillings has declined consistently from 0,42 per cent in 1999 to 0,26 per cent in 2007 (Kaplan, 2009).

Table 1.2 and Figure 1.2 show the number of new trademarks, patents, designs and copyright in films registered and granted by CIPRO in 2007/08 and 2008/09.

Table 1.1: Patent applications of South African origin filed through the PCT, 1994 to 2007 (Kaplan, 2009. WIPO Statistics Database and USPTO)

	'94	*'95*	*'96*	*'97*	*'98*	*'99*	*'00*	*'01*	*'02*	*'03*	*'04*	*'05*	*'06*	*'07*
South African Patents	30	42	72	84	114	317	387	419	384	357	411	358	422	360
USPTO grants	109	127	116	114	132	127	125	137	123	131	115	108	127	116

Table 1.2: CIPRO's Record of South African IPR

	2007/08	*2008/09*
Trademarks	31 627	28 671
Patents	10 830	12 794
New designs	2 298	2 2511
Copyright in films	104	51

Table 1.3: Granted Plant Breeder's Rights (Source: South African Agricultural Research Council Data)

Year	*'94*	*'95*	*'96*	*'97*	*'98*	*'99*	*'00*	*'01*	*'02*	*'03*	*'04*	*'05*	*'06*	*'07*	*'08*	*'09*
Number of Granted PBR's	6	17	10	18	27	4	20	10	8	17	20	21	11	5	20	28

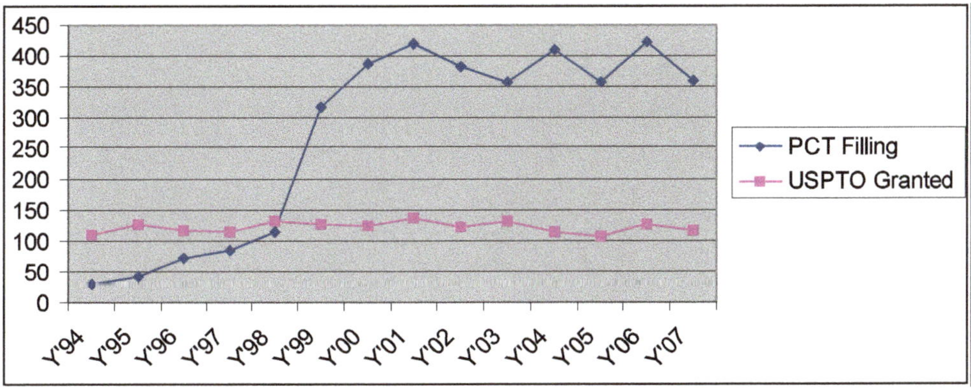

Figure 1.1: Patent Applications of South African Origin Filed through the PCT, 1994 to 2007 (Kaplan, 2009, sourced from WIPO Statistics Database and USPTO)

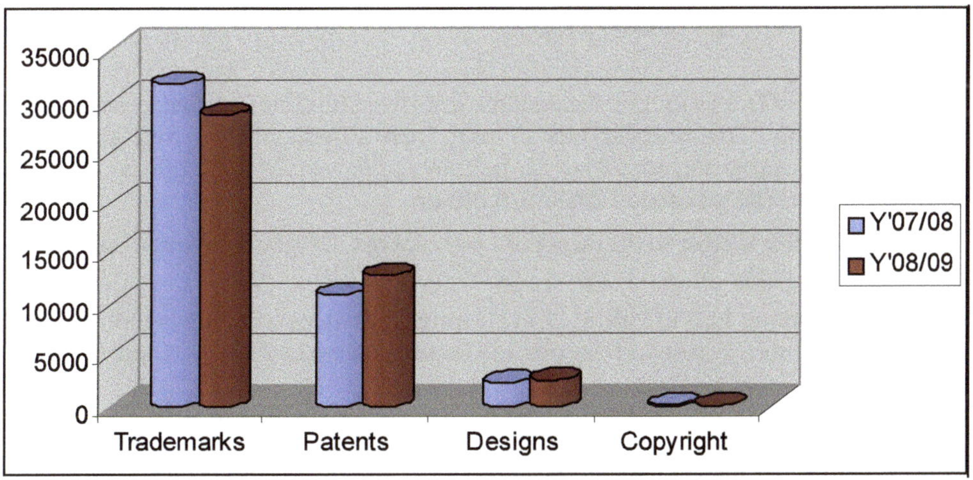

Figure 1.2: CIPRO's Record of South African Intellectual Property Rights

Table 1.3 and Figure 1.3 show the number of granted Plant Breeder's Rights (PBR's) through the South African Department of Forestry and Fisheries.

South Africa's Gross Expenditure on Research and Development (GERD) as a Percentage of the Gross Domestic Product (GDP)

GERD expressed as a percentage of GDP provides an indication of the concentration or intensity of R&D in an economy. There has been a steady increase in GERD as a percentage of GDP from 0,60 per cent in 1997 to 0,95 per cent in 2006. The Figure 1.4 show that GERD as a percentage of GDP decreased slightly from 0,95 per cent in 2006/07 to 0,93 per cent in 2007/08 (South African National R&D Surveys and Statistics South Africa P0441 Gross Domestic Product, First Quarter 2008).

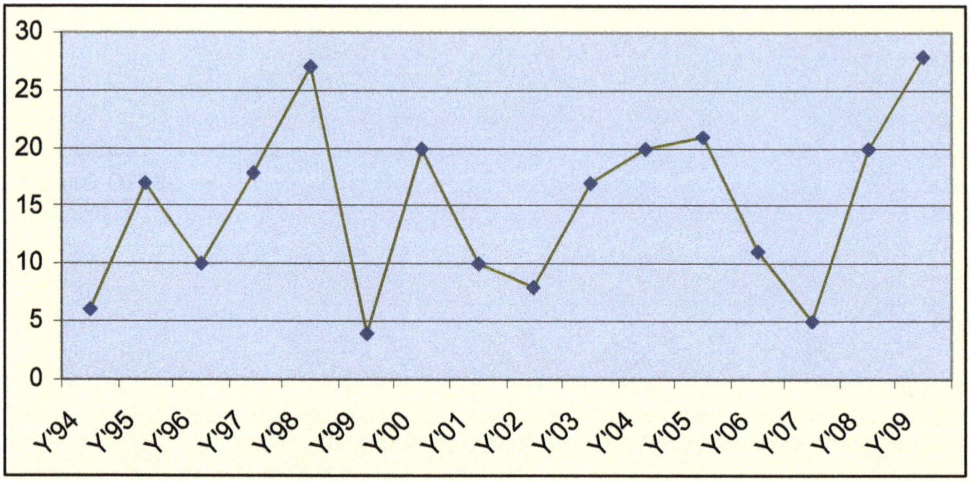

Figure 1.3: Granted Plant Breeder's Rights

Over the past five years, government has improved funding for science, technology and innovation. The budget has grown from ZAR2 billion in 2005/06 to ZAR5, 1 billion projected for 2011/12. In 2009/10 it was ZAR4, 2 billion. In 2006, South Africa's GERD exceeded ZAR16, 5 billion.

Although this is close to the target of 1 per cent of GDP, the investment is modest; it is time the target increased beyond 1 per cent of GDP funding for research.

Table 1.4 shows in 2007/08, 31 352 full-time equivalent personnel were engaged in R&D. There are 1,5 researchers per 1 000 employees (in 2004/05 there were 2,7

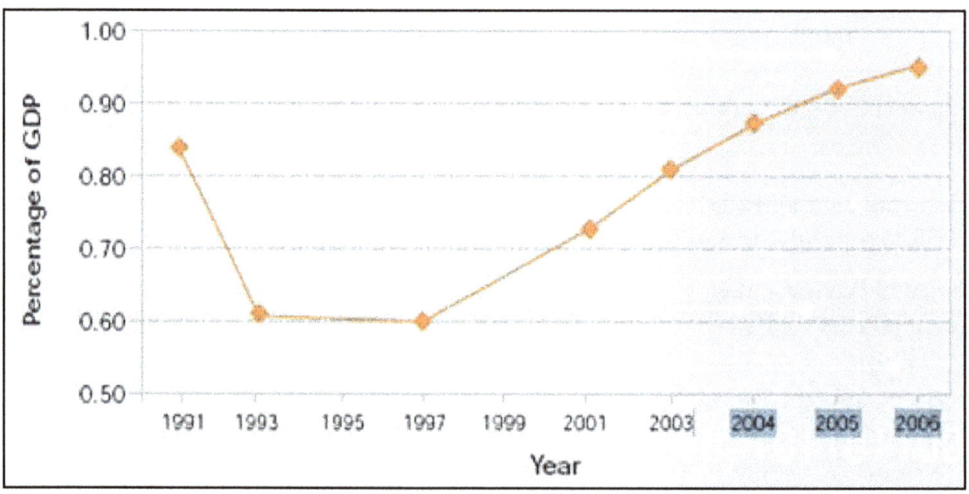

Figure 1.4: Gross Expenditure on R&D as a Percentage of Gross Domestic Product (South African National R&D Surveys and Statistics South Africa P0441 Gross Domestic Product, First Quarter 2008)

researchers per 1 000 employees), which is low relative to industrialised countries. In 2007/08 the number of full-time equivalent researchers increased by around 748 people.

Table 1.4: Key Figures (South African National R&D Surveys and Statistics South Africa P0441 Gross Domestic Product, First Quarter 2009)

Indicator	*Value*	*Value*
Gross expenditure on R&D–GERD (Rand Millions)	2006/07	2007/08
Gross domestic product (GDP) at market prices (Rand Million)	16520.6	18624
GERD as a percentage of GDP	0.95	0.93
Basic research as a percentage of GDP	0.18	0.19
Total R&D personnel (FTE)a	30986	31352
Total researchers (FTE)b	18572	19320
Total researchers per 1000 total employment (FTE)c	1.5	1.5
Total R&D personnel per 1000 total employment (FTE)	2.5	2.4
Civil GERD as a percentage of GDP	0.89	0.87
Total researchers (headcount)	39591	40084
Women researchers as a percentage of total researchers	39.7	40.3

a: FTE: Full Time Equivalent

b: Following OECD practice, doctoral students are included as researchers.

c: Total employment is provided by the OECD from the International Labour Organisation based on the Labour Force Surveys of Statistics South Africa.

Sources: South African National R&D Surveys and Statistics South Africa P0441 Gross Domestic Product, First Quarter 2009.

International Property Rights Index Report

South Africa was ranked 24th in a study that shows that countries that protect the physical and intellectual property of their people enjoy nearly nine times higher GDP per capita than countries ranking lowest in property rights protection. The 2009 International Property Rights Index (IPRI) compared the protection of physical and intellectual property to economic stability in 115 countries representing 96 per cent of the world's GDP.

The underlying assumption of the study of IPRI is the correlation between having an effective property rights regime and its significant impact on the economic performance of a country.

The IPRI is a composite ranking of three comprehensive areas of property rights, namely the Legal and Political Environment, Physical Property Rights, and IPR. Of the 115 countries included, the top quartile averaged US$39,991 in GDP per capita while the average in the bottom 20 per cent was only US$4,341 per capita. The second, third and fourth quartiles averaged US$23,982, US$11,748, and US$4,891 respectively.

The nearly linear data trend shows that countries placing a high priority on property rights see increased economic security.

South Africa's overall score declined from 6,9 in 2008 to 6,8 in 2009 and remained at 6,8 in 2010 (out of 10 with 10 being the strongest score and 0 the weakest). It has moved from the top quintile to the second-highest.

The score for IPR improved from 7 (2008) to 7,4 (in 2009 and 2010), which is a creditable improvement, while that for the Legal and Political Environment declined from 6,6 in 2008, to 5,9 in 2009 and further declined to 5,7 in 2010.

Legal and Political Environment

The Legal and Political Environment component represents the fundamental essence of a free environment in which judicial independence and the rule of law bestow on individuals the right to benefit from the existence of a strong private property rights system. A positive combination of judicial independence, the rule of law, political stability and the control of corruption creates an environment that is conducive to the growth of the economy of the country through a system which upholds the protection of property rights. Under this component, judicial independence, the application of the rule of law, political stability and the control of corruption in the country are measured.

Physical Property Rights

A strong property rights regime advocates the protection of private property rights and ensures the implementation of policies, procedures and laws concerning the registration of property and creates an opportunity for individuals and businesses to access credit thereby providing an opportunity to actively participate in economic movement. South Africa's 2010 PPR score is 7,4.

Intellectual Property Rights

The IPR component evaluates the protection of intellectual property covering all areas of IPR; additionally it reviews the country's policies and their effectiveness in enforcing patents and copyrights. Table 1.5 and Figure 1.5 show the South Africa's IPR Score for 2010 is 7,4.

It is encouraging that South Africa's IPR is ranked at 7,4 which compares well with regions such as North America and Western Europe.

Table 1.5: South Africa's IPRI Ranking 2007 to 2010

Year	*IPRI*	*LP*	*PPR*	*IPR*	*World ranking*	*Quintile*
2007	6,4	5,8	6,6	6,8	21 (out of 70)	2nd
2008	6,9	6,6	7,1	7,0	24 (out of 115)	1st
2009	6,8	5,9	7,1	7,4	28 (out of 115)	2nd
2010	6,8	5,7	7,4	7,4	24 (out of 125)	2nd

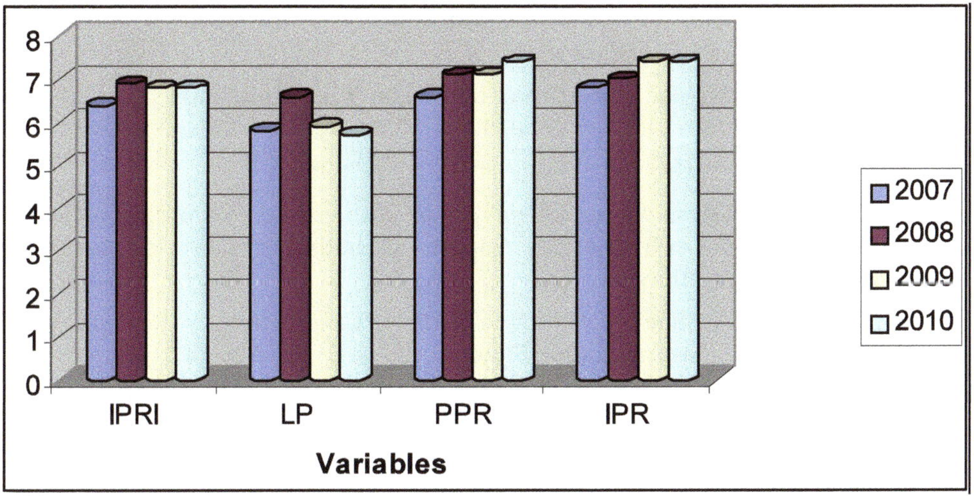

Figure 1.5: South Africa's IPRI Ranking 2007–2010

Conclusion

Various policy initiatives largely driven by the DST and its instruments have played and will continue to play an important role in the institutions' contribution to South Africa's economic growth. IPR, in terms of patenting activity, depends on research activity and research output. In recognition of the lack of capacity in the area of IP management, the Innovation Fund set up a patent attorney development programme aimed at addressing the racial imbalance within the South African IP system while simultaneously contributing to human resource development for enhanced IP management and commercialisation.

References

http://www.bowman.co.za/IntellectualProperty/PlantBreedersRights.asp Visited 24/06/2010.

http://www.cipro.org.za Visited 07/06/2010.

http://www.engineeringnews.co.za/article/sas-rd-spend-as-percentage-of-gdp-slowing-pandor-2009-12-07 Visited 08/06/2010.

http://www.gcis.gov.za/resource_centre/sa_info/pocketguide/2009/018_science_and_technology.pdf Visited 08/06/2010.

http://www.hahn.co.za/patent-services.htm Visited 07/06/2010.

http://www.saiipl.org.za/introductio-copyright.htm Visited 07/06/2010.

https://www.sabs.co.za/content/uploads/files/IP_Guide_FA_web.pdf Visited 08/06/2010.

http://www.property24.com/articles/sa-24th-in-property-rights-index/9295 Visited 09/06/2010.

South African National R&D Surveys and Statistics South Africa P0441 Gross Domestic Product, First Quarter 2008.

The Department of Science and Technology's Ten Year Innovation Plan (2008 to 2010).

Intellectual Property Rights Act, 2008 (Act No. 51 of 2008).

Kaplan, D., 2009. *Intellectual Property Rights and Innovation in South Africa: A Framework* in The Economics of Intellectual Property in South Africa. World Intellectual Property Organisation.

Sibanda, M., 2007. *The State of patenting in South Africa – Analysis of the South Africa patent landscape*. The Innovation Fund.

Sibanda, M., 2009. *Intellectual Property, Commercialisation and Institutional Arrangements at South African Publicly Financed Research Institutions* in The Economics of Intellectual Property in South Africa. World Intellectual Property Organisation.

Chapter 2
Intellectual Property (IP) in Botswana: Research but No Patents

Botshelo Maedza
Acting Principal Civil Engineer,
Botswana Technology Centre, MBIE
E-mail: har709@botec.bw

ABSTRACT

In Botswana, Section 11 (4) of the Industrial Property Act of 1996 states that in the absence of any agreement the Intellectual Property created by an employee belongs to the employer. In the absence of Intellectual Property policy in Research and Technology Organizations and Universities, can the outcome of research be fully realized? Can researchers in these institutions be innovative when they know that they will not reap the benefits of their Intellectual Property? There seem to be lack of intellectual property awareness, generally. Could this be the reason why research has been conducted for many years but no patents have yet been registered? Botswana Technology Centre (BOTEC), a wholly government funded institution, despite celebrating 30 years of existence in 2009, is yet to register its first patent. The same applies to Rural Industries Innovation Centre, a subsidiary of Rural Industries Promotions Company (Botswana), which is also government funded, produces several industrial and agricultural implements and equipment that have changed many people's lives. But none of them has be protected. The University of Botswana conducts a lot of research, but its research is geared towards publications. This chapter looks at why despite the government of Botswana investing substantial money every year in research, very little has been achieved in terms of producing Intellectual Property. The paper will focus on Botswana's overall policy framework, to distinguish whether the policy makes an effort to create an innovative environment for knowledge workers and other people. The paper also makes a case study of Botswana Technology Centre and reviews why the institution has no IP policy despite being more than 30 years in existence.

***Keywords**: Intellectual property, Patent, Innovation, Science and Technology, Capacity building.*

Introduction

Botswana spends about 0.4 per cent of its Gross Domestic Product (GDP) on Research and Development (R&D) which is below 1 per cent, the minimum requirement for emerging innovative economies. However, the question arises: Is this budget too little to justify the non contribution of Research, Science and Technology in Botswana?

Botswana has a Science and Technology (S&T) policy which was approved by Parliament in July 1998. The policy resulted from a concerted effort by the Government of Botswana and the Botswana Technology Centre (BOTEC) to develop a national S&T landscape. The primary purpose of the new policy was to be used by the Government as guidelines for the future development of S&T in the country.

However, despite the policy clearly highlights the areas of S&T needed to be developed; the implementation of the policy by the Government has made very little progress to realize its objectives. Research and Technology Organizations (RTOs) such as BOTEC and Rural Industries Promotions Company (Botswana) (RIPCO (B)), and the only university in the country, currently, The University of Botswana (UB), though conducting a lot of research, has very little to show as the outcomes of their research in S&T. In fact, BOTEC and RIPCO (B) up to now, have no Intellectual Property (IP) Policies, twelve years after the existence of the S&T policy and more than thirty (30) years since their formation. Can the two organizations be realistically expected to produce any IP in S&T when they do not have IP policies in place?

Several other initiatives have been put in place by the government to try and improve the S&T landscape in Botswana. These include: the development of Botswana National Research, Science and Technology Plan of 2005, establishment of Botswana International University of Science and Technology (BIUST) (under construction), identification of national priority in S&T areas, establishment of Botswana Innovation Hub and establishment of Botswana Research Science and Technology Funding Bill. Despite all these deliberate efforts by government, very little progress has been made with regards to contribution to the economy by research institutions. Are the initiatives above implementable or is it failure on the part of research institutions to implement such initiatives?

The paper forms the preliminary work which is intended to form part of a study which will be conducted as part of a Ph.D. degree proposal which the author wants to undertake. Consequently, the paper poses more questions than try and offer answers, and communicates the perceived challenges that affect the development of IP in Botswana.

This chapter lays the groundwork for research on why Botswana, despite being involved in extensive research on various areas of S&T, is unable to develop its IP to enable the research institutions, which happen to be funded by government, realize the benefits of their investment.

Understanding of Intellectual Property

Intellectual property can be defined as the legal rights which result from intellectual activity in the industrial, scientific, literary and artistic fields. Once an idea is conceived, it goes through various stages of development. These stages are critical in development of IP. It is imperative to separate a research activity from IP development. This is because with an IP policy in place, it should be clear how the two correlate.

Every research activity is expected to yield results and if these results are positive, then some form of IP must be registered from the research activity. Consequently, the research scientist must know basic principles that regard IP. For example, the scientist must understand issues pertaining to proper documentation of manuscripts and protection of information to prevent disclosure to the public. Once the research information is disclosed to the public it can no longer be protected by, say patent.

For an organisation doing research it is critical to have an IP policy because all parties involved in research, would understand clearly their roles and responsibilities guided by the IP policy.

Intellectual Property Development in Botswana

Botswana has a population of nearly 2 Million and a GDP of US$7,032 (2010 estimated) (Source: Capital Finance). With the economy heavily depended on natural resources, minerals in particular, Botswana has serious challenges to start diversifying the economy. Even though the desire to diversify the economy has been there for many years, very little has been achieved on the ground, because minerals, a non-renewable resource, are still the mainstay of Botswana economy.

S&T has been identified as the potential driver in the diversification of the economy. As a result the government has put in place, several initiatives to try and address the issue of over-reliance in natural resources and encourage exploitation of alternative income streams. These initiatives include:

1. *Botswana Science and Technology Policy of 1996* (under review): The main objective of this Policy, which is now undergoing review, is to recognize the role that S&T could play in diversifying the economy. The policy explains how the increased capacity in S&T would result in improvements in national economy competitiveness and the elevation of the standard and quality of life of Botswana.
2. *National Research, Science and Technology Plan of 2005*: This Plan identifies Botswana's priority areas for investment in research and presents an implementation programme which is aligned with National Development Plans. The Plan identifies key issues such as:
   - ☆ Listing of priority areas for investment in research;

- ☆ Identifying human resources and institutional structures required to implement the plan (including the formation of new funding and research agencies); and,
- ☆ Developing an implementation plan with estimated financial costs.

This plan also identified several actions that are necessary for its effective implementation. These include provision of adequate funding, appropriate and efficient institutions, the establishment of plan monitoring mechanisms, as well as train effective research staff.

Under these actions:

- ☆ The government would be required to commit to a research expenditure of 1 per cent of GDP.
- ☆ Research funding would largely be allocated on a competitive basis. This is aimed at encouraging more relevant and efficient research focusing on Botswana's priority development needs.
- ☆ Botswana Research, Science and Technology Funding Agency would be established as a mechanism to manage competitive funding.
- ☆ Rationalization of government institutions such as merging the National Food Research and Technology Centre (NAFTEC) with the Department of Agricultural Research into a National Food and Agricultural Research Centre; and Botswana Technology Centre with Rural Industries Promotions Company (Botswana) to form new institution would be undertaken.

3. *Industrial Property Act of 1996*: Botswana has an Industrial Property Act. The Act provides for the registration of patents, trademarks, designs and utility models. The Act offers internationally accepted standards of protection and provide the Government with the means to regulate the Intellectual Property Rights protection system and penalize against infringement.

   According to one of the Industrial Property Officers at the Registrar Companies under the Ministry of Trade and Industry, the office has not been able to build capacity in terms of training their officers to equip them with skills which will enable them to execute their mandate effectively. The office has not discharged its mandate of educating the nation about IP issues. This is due to lack of placing the needs of the department at the top of government priority list. The IP training that is taking place at the department is too little to effect significant change in both short and long term.

Developmental and Management Approach to IP – The Case of BOTEC

BOTEC or indeed Botswana, is lacking behind in terms of capacity building in the area of IP, hence the contribution by IP to the economy is insignificant. A drastic change needs to be made in order for BOTEC to begin generating research results and exploit the outcome of their work. This could be addressed by focusing on the following:

1. Developing, adopting and effectively implementing a rigorous IP policy to provide guidance for the management and exploitation of IP.
2. Lobby for a policy and legislative regulatory framework that ensures good practice and one that will promote effective transfer of technology and consequently attract sponsors, financiers or partners.
3. Create an environment conducive for research and one that will stimulate creativity and innovation and ensure fair and equitable sharing of benefits.
4. Move away from being wholly funded by government and start sourcing or competing for funding from donors and international agencies. Alternative funding resources would help pay for IP registration costs (which are also quite high).
5. Develop increased network between institutions and across sectors.
6. Embark on extensive human capital development in both scientists and support staff.
7. Establish specialized IP and commercialization support. This could be by establishing a fully fledged IPR office that will assist researchers in patenting their inventions and creations, and protect their rights so that they could benefit materially from their work.

   Financiers like to invest in patentable projects, therefore only IPR office would be in a position to assess and advise whether or not the research outcome is patentable.
8. Establish structures that provide for an enabling environment for intellectual property creation, protection, management and commercialisation. At the moment BOTEC does not have commercialisation policy which guides the institution on commercialisation activities. As a result products are being developed with knowing how they would be commercialised, and indeed several products have ended up not effectively commercialised.

Government needs to provide a funding agency where institutions will compete for funding based upon the relevance of their proposal. All these initiatives should be part of the government's larger goal aimed at driving Botswana towards a knowledge-based economy, in which the production and dissemination of knowledge leads to economic benefits.

Analysts predict that effective IP strategies can raise African producers' incomes by up to 300 per cent. A study, "Distinctive Values in African Exports: How Intellectual Property can raise export income and alleviate poverty," conducted by Light Years IP, in 2008, and funded by UK Department of International Development, showed that IP is increasingly being used in business strategies by African producers to boost their export income. From the 14 product studied sectors, it was found that these sectors have the potential to increase export income from US$1.1-billion per year, to between \$2.5-billion and \$3.5-billion per year. The report further states that if IP was "applied properly to sub-Saharan exports, which earned an estimated \$9-billion per year, these strategies would increase the export income in this region to between \$20-billion and \$27-billion per year."

Conclusion

Even though the government of Botswana came up with relevant policies and legislation; the biggest challenge that the government faces is capacity building. Lack of people adequately equipped with skills and expertise to implement the policies is highly problematic and it will take some time to reverse the status. Government trains people, but those skilled people leave the country for better opportunities elsewhere resulting in the "brain drain" phenomenon. In order to curb that and attract better skilled workforce to the country and retain existing talent and experts, the government has recently introduced scarce skills allowance for skills that are considered to be scarce in the economy.

Accountability seems to be another serious problem that the government has to deal with. Government funded institutions should be made to account for the funds they get from the government. Return on investment should be the norm at government funded institutions. Institutions should report on progress they have made in terms of technologies they have produced, experts or scientists they have trained and those who are undergoing training, whether formal or on-the-job training. This would help monitor progress that the government is making by funding research institutions.

Acknowledgements

The author is grateful of contributions made by Mr. Oabona C. Monngakgotla, Principal Research Science and Technology Officer, Department of Research Science and Technology (Intellectual Property Rights), Ministry of Infrastructure Science and Technology, Botswana; and Mr. Ikanyeng Moshabi – Industrial Property Officer, Registrar of Companies, Ministry of Trade and Industry, Botswana.

References

Botswana National Research Science and Technology Plan – Final Report (2005). Centre for Scientific and Industrial Research, South Africa. Prepared for Ministry of Communications, Science and Technology, Botswana.

Distinctive Values in African Exports: How Intellectual Property Can Raise Export Income and Alleviate Poverty (2008). Light Years IP. Funded by UK Department of International Development.

Science and Technology Policy for Botswana (1998). Government of Botswana.

Chapter 3
Intellectual Property Rights: Situation in Botswana

Letsomo la Tshipa
Performing Arts Officer in the Department of Arts and Culture,
Ministry of Youth, Sports and Culture,
Botswana
E-mail: itshipa@gov.bw

ABSTRACT

Intellectual Property Rights matters are a new phenomenon to Botswana. It therefore becomes obvious that the status does not afford Botswana opportunity to tap the economic, social and even cultural benefits derived from the Intellectual Property Copyrights. The level of awareness on Copyright in Botswana is very low but the interest is great. Here is a brief write up on the prevailing situation of Intellectual Property Rights in Botswana.

***Keywords**: Intellectual Property Rights, Copyrights, counterfeit*

Introduction

Intellectual Property Rights matters are a new phenomenon to Botswana and other developing countries. It therefore becomes obvious that the status does not afford Botswana opportunity to tap the economic, social and even cultural benefits derived from the copyrights. The level of awareness on Copyright in Botswana is very low but the interest is great.

Background

In order to give the desired attention to copyrights matters, the Botswana government has established the Copyright Office in Botswana. This Copyright office was established under Section 22A of the Copyright and Neighboring Rights Act. This office is established under the Ministry of Trade and Industry in the Department of Registrar of Companies and Intellectual Property. This office is responsibility for organizing training workshops for the creators and other stakeholders in collaboration with other relevant stakeholders like the Department of Arts and Culture, and the University of Botswana.

Establishment of Copyrights Society of Botswana (Cosbots)

Botswana is currently finalizing the process of establishing the Copyright Society of Botswana through the Department of Registrar of Companies and Intellectual Property. The society; which will be a collecting society and it was expected to be operational before the end of the year (2010). It is going to be provided with start up capital by the government through the levy accrued from the fund for blank tape and other recording devices.

It is encouraging to note that the Government of Botswana has reviewed the Copyrights Act of 1965 which was inherited from the Colonial administration. This review which was done 1998 was necessitated by the fact that the old act could no longer address Copyrights issues due to new developments especially technological changes.

Awareness on Copyrights

The level of awareness on Copyright in Botswana is very low, but the interest is great. The Department of Arts and Culture provides courses related to Copyrights utilising international experts on Copyrights to lecture on the related issues. The Department of Arts and Culture also have done training workshop on copyrights specifically for law enforcement officers. This was done as a result of artists registering complaints with the Department of Arts and Culture about the counterfeiting or pirating of their art works.

Levy On Technical Devices Fund

As a move to strengthen the Copyrights case in Botswana, the Government of Botswana through the Ministry of Trade and Industry has established the above mentioned Fund. This imposed a levy on all imported and locally produced blank sound and audio visual carriers, compact discs and equipment capable of being used to copy protected created material. This will be collected through the Ministry of Finance by Botswana Unified Revenue Services (BURS). This fund will be used to fund project that are geared at the development of Copyrights holders. This fund will also be used to support projects that are aimed at educating, informing and communicating copyrights issues to the public.

Conclusions

While the awareness regarding Intellectual Property Rights matters in Botswana is still at a juvenile stage, but the Government is putting a lot of efforts in this direction to harness socio-economic benefits for the people by putting into action various schemes and methods.

Chapter 4

Country Report on IP Policy and IP Related Laws

Petros Orton Shiyombo Musukwa
Technology Transfer Officer
National Commission for Science and Technology
P/Bag -B303, Lilongwe 3, Malawi
E-mail: pmusukwa@ncst.mw, ptrsmusukwa@yahoo.co.uk

ABSTRACT

The basic premise of Intellectual Property Policy is that Intellectual Property is closely connected to socio-economic and technological development. There is a symbiotic relationship between IP and development in that it improves the quality of life through technological advancements and enjoyment of literary and artistic works. A clear and comprehensive policy on the IP system is therefore critical in the effective implementation of government's development agenda. Malawi has however realized the importance of IP Policy and other IP related laws for her to reposition herself in the competitive global economy.

***Keywords**: Intellectual Property, Policy, Socio-economic.*

Introduction

In order to exploit the opportunities in the intellectual property related issues for socio-economic and technological development, there is a need to build an effective intellectual (IP) system for the Country. Therefore by reviewing old IP Laws and updating those that would accommodate the dynamic global economies, Malawi will have effective IP laws that will stimulate research, technology development and innovation. Strong IP system will set a strong base for the effective implementation of government's development agenda.

Without IP system in place, Malawi is not in a position to effectively implement its IP related development policies, is not competitive in the global economy and has not integrated IP issues in the current government development policies, especially in the Malawi Growth and Development Strategy (MGDS) and in other sector Policies contributing to the achievement of the (MGDS). These sector Policies are:

1. The Vision 2020 (1998 which envisages Malawi as a technologically driven middle-income economy

2. A number of National policies in the sectors of Agriculture, Health, Energy, Environment and Natural Resources, Education, Security, Science and Technology among others.

In the context of the phenomenon of globalization, this is a serious setback for the country.

Industrial Property Laws in Malawi

However, Malawi has realized the importance of IP Policy and other IP related laws for her to reposition herself in the competitive global economy. Therefore the country has developed the National IP Policy awaiting approval by Cabinet and is reviewing the following outdated IP Laws that have been in use for sometime:

- Patents Act (Cap. 49: 02), 1958, revised 1985
- Copyright Act (Cap. 49: 03), 1989
- The Trade Marks Act (Cap. 49: 01), 1958
- Trade Descriptions Act (Cap. 49: 04), promulgated in 1987 basically to criminalize certain uses of trade descriptions and trade mark forgery
- Registered Designs Act (Cap. 49: 05) 1958, revised 1985
- Competition and Fair Trading Act (1989) :The Competition and Fair Trading Act was enacted in 1989 to regulate and monitor monopolies and concentrations of economic power as well as prohibit anti-competitive trade practices
- Environment Management Act (Caps. 48:09 and 60:02). The Environment Management Act, (1996) is in process of being replaced by another statute to take into account the protection of genetic resources.
- The Plant Varieties Protection Bill, 2006, introduces the protection of new varieties of plants.

Adherence to International Conventions

Malawi has the potential to adhere to the IP issues because it has been linked to the following international agreements and treaties:

- Malawi is member of the World Trade Organization (WTO), World Intellectual Property Organization (WIPO) and United Nations Educational, Scientific and Cultural Organization (UNESCO).

- Among the WTO agreements that bind Malawi, the most important in the field of IP is the Agreement on Trade-Related Aspects of Intellectual Property Rights (TRIPS).
- In the field of copyright and related rights Malawi is also a party to the Universal Copyright Convention which is administered by UNESCO.
- An important dimension to Malawi's IP regime has been added by the country's adherence to the Convention on Biological Diversity (CBD) adopted in 1992.
- At regional level, Malawi is a founding member of ARIPO. Malawi is a party to ARIPO's Protocol of Patents and Industrial Designs (1984) and the Banjul Protocol on Marks (1993).
- Malawi is also a founding member Common Market for Eastern and Southern Africa (COMESA) and the Southern Africa Development Community (SADC).
- COMESA Treaty obliges member states to *"jointly develop and implement suitable patent laws and industrial licensing systems for the protection of industrial property rights and encourage the effective use of technological information contained in patents."*
- Article 24 of the Protocol on Trade in SADC, states that *"Member States shall adopt policies and implement measures within the Community for the protection of Intellectual Property Rights, in accordance with WTO Agreement on Trade-Related Aspects of Intellectual Property Rights."* The SADC Protocol on Culture, Information and Sports also calls on Member States to provide effective copyright and related rights protection and to fight against piracy.
- At continental level, the African Union Heads of State and Government have decided to establish the Pan African Intellectual Property Organization.

Conclusions

In conclusion, for Malawi to effectively implement her IP Policy, proposes that the following strategies should be considered:

- Establish the National IPR and Policies priorities
- Review the Current IPR Laws
- Establish critical/important gaps between Malawi, regional and international Agreements
- Propose ways for harmonisation of IPR and policy adoption
- Establish Intellectual Property Office or patent office.
- Malawi needs to ratify to important International Agreements on IP

Chapter 5

Zambia: Status Paper on Intellectual Property

Thompson Namukaba
C/o National Arts Council of Zambia
P.O. Box (10101) 50812, Lusaka, Zambia
E-mail: thompsonnamukaba@yahoo.com, namukaba@gmail.com, nacz@coppernet.zm

ABSTRACT

With a population of well over 11 million, Zambia's Industrialization is slowly gaining momentum. Recent investment opportunities for both local and foreign investors have given rise to a number of different IP based products, produced locally which later find their way on local and international markets.

Zambia has 72 language groups in its nine provinces that are rich in numeracy, cultural activities as well as an abundance of traditional knowledge and Arts.

Zambia 's Research and Development Institutions have their stories to tell in terms of a luck of IP policy to help support the industry's endeavours in coming up with new solutions to technical problems faced by the world today. University of Zambia Intellectual Property Policy Page: Introductory remarks.

There is much to be desired of Zambia; a lot of untapped potential in traditional knowledge and the Arts, which if harnessed properly could set a record on the nation's GDP.

Geographical indications and cultural heritage;

Specialty of Zambia's land in certain regions and the rich cultural heritage, gives the country unique exploitable business opportunities for locals and foreign investors which have not been fully exploited to gain leverage on the wider market worldwide. Not to mention a lot of unpatented plant varieties as well as inventions mainly in the form of traditional medicine, the matter which is currently being addressed under the Ministry of Health.

The ambiguity of who owns what *right* under the cultural heritage's many facets is one factor that evidently tells the need for IP awareness amongst countrymen and women. Such an extraordinary phenomena may well need an extraordinary approach coupled with IP knowledge of the subject for better management of IPR.

Keywords: *Creation, Development, Promotion, Exploitation, Protection, Trade, Innovation.*

Introduction

Intellectual Property is still an abstract subject not only to a number of users but to creators of IP works as well; there is need for more sensitization programs on IP to reach the desired level of understanding the subject matter. Zambia has, along with the world suffered adversity brought about mismanagement of IPR. Many investors have taken advantage of high poverty levels to supply counterfeit products at relatively cheaper prices and have on many occasions gotten away with it.

The UNZA IP seek to influence formulation of sound IP policies to better manage IPR in the nation and to spearhead the integration of IP studies in the national curriculum for the subject to become a common language.

An achievement of this kind will entail an IP cultured nation ready to achieve more in its ventures towards development.

Intellectual Property Laws

Zambia is a signatory to the Berne convention and a number of other world conventions on IP laws, some of which, though yet to be reviewed due to advancement in technology, are well stipulated in the country laws with *little* or *no* awareness to the creators and users of IP based products.

For example, Copyright and Performance Rights ACT # 44 of 1994, Provides for protection of copyright protected materials and related rights with exceptions and limitations.

IP Administrative Bodies

Zambia has two major IP administrating Institutions:

PACRO, manages the filing of patents and registration of Companies under the Ministry of Commerce, Trade and Industry.

Copyright Registrar under MIBS, is the custodian of the copyright Act and supervises copyright licensing organizations like CMOs. The office of the Registrar of copyright is also highly committed to ensure IP compliance in the nation. It has conducted numerous raids on infringing equipment and goods within Zambia and at points of entry with the help of the customs at clearance border points (Figure 5.1). To help manage this tasking venture, the office is working closely with Zambia Bureau of Standards and Zambia police, in tracing the culprits and bringing them to book.

Zambia has a number of IP infringing cases. Most recently the Intellectual Property Unit under Zambia police raided an undercover company which produced

Figure 5.1: Pirated DVDs packed in a bag of rice seized by the Customs Duty Officers

counterfeit products ranging from Johnson and Johnson's baby powder to petroleum jelly, body lotions, bathing soaps, biscuits and sweets and supplied the local market undetected for many years. Another undercover Company which reproduced DVDs and audio CDs without a license is currently facing the law.

In Zambia it is not feasible to successfully get rid of counterfeit products because almost all goods sold are counterfeit. The sellers and buyers know they are selling or buying counterfeit products and it's openly discussed.

Infringing Activities

The Zambian Market is highly diluted with infringing goods of almost all consumables approximately 90 per cent of these in form of clothes, electrical appliances, drugs to mention only a few.

A number of infringing and health hazard goods are on many occasions seized by customs officers at points of entry into the country and destroyed, as many more enter undetected.

The Creative Industry and Intellectual Property

National Arts Council of Zambia (NACZ)

All Arts activities fall under the umbrella body of NACZ. The growth of the Creative Industry (being composed of creators) is highly threatened in the absence of IPR management and as such a venture would paralyze innovation and creativity of the people.

Example; VAC an affiliate of NACZ

- ☆ VAC has branches (galleries) countrywide with Henry Tayali gallery in Lusaka being the head office.
- ☆ VAC organizes and conducts exhibitions on behalf of Artists while retaining a percentage on each sale for administrative purposes.
- ☆ VAC also holds training and interactive workshops for Artists to help expose local talent locally and internationally for a wider market.

For such an ambitious organization knowledge of IP is a necessity for better trade.

The Arts

The highest qualification level attracts a diploma at a locally renowned Public Arts Institution (Evelyn Hone College) based in Lusaka. However, a recently opened University (Open University of Zambia) offers Degree programs in fine Arts. A good IPR management system will ensure a healthy Art industry.

Collective Management Organisations (CMO)

Only two CMOs exist in Zambia

1. ZAMCOPS for Musicians and phonogram producers
2. ZARRSO

ZAMCOPS which licenses users of Musical works started in 1996

In 2008 it collected well over US$ 119,025 (K 513,000,000.00)

ZARRSO for literary and Artistic works was launched in 2008

Rights holders include Visual Artists, Writers, Publishers and journalists.

Licensing Contracts

Assignment and Licensing

The rights holder may authorize another through a license to exercise their economic rights.

License can be granted for any work that qualifies for protection.

Exceptions and Limitations (Striking a balance)

The severity and flexibility of rights conferred sometimes have to be balanced by means of exceptions and limitations. The copyright law provides for exceptions and limitations for;

- ☆ The purposes of research or private use and for educational purposes.
- ☆ Use of works for parliamentary or judicial proceedings or inquiries.
- ☆ Public awareness, and news reporting, provided the author or source of information is mentioned.

Challenges to IP Development

- Lack of political and business will to develop and enforce IP laws
- Political relegation of most IP matters
- Non or leftover budget allocation
- No national IP strategy
- Disrespect for IPR due to perceived necessity to copy
- Lack of awareness
- Executive ignorance
- Lack of capacity

Way Forward

- Mass appeal
- Riding the wave of curiosity
- Exhibitions, IP expos and Road shows
- Outreach and establishing media presence
- The power of synergy. Make use of everyone

Regional Commitment

- SADC Protocol on IP
- AU IP desk
- IP ambassador for Africa
- Observer status for WIPO at all SADC and AU summits

Conclusion

A series of sensitization workshops and seminars conducted by WIPO, IFRRO, Kopinor and CISAC has created a formidable stance marking a promising beginning in IPR management. The stance however, is highly dependent on concerted efforts in all sectors of human endeavor, the government, private sector and individuals. Therefore, we need effective IP policies as well as dedication and energy to finish what we have started to ensure steady development.

Aknowledgements

Mr. Kenneth Musamvu Copyright Registrar (MIBS)

Mr. Kingsley Nkonde Intellectual Property Specialist

Mr. Victor Makashi NACZ Director

References

Copyright Act No: 44 of the laws of Zambia.

University of Zambia Intellectual Property Policy.

Abbreviations

AU:	African Union
CISAC:	International Confederation of Societies of Authors and Composers
CMO:	Collective Management Organization
GDP:	Gross Domestic Products
IFRRO:	International Federation for Reprographic Rights Organization
IP:	Intellectual Property
IPR:	Intellectual Property Right
MIBS:	Ministry of Information and Broadcasting Services
NACZ:	National Arts Council of Zambia
PACRO:	Patents and Companies Registration Office
UNZA IP:	University of Zambia Intellectual Property
WIPO:	World Intellectual Property Organization
SADC:	Southern Africa Development Community
VAC:	Visual Arts Council
ZAMCOPS:	Zambia Music Copyright Society
ZARRSO:	Zambia Reprographic Rights Society

Chapter 6

Intellectual Property Management in Zambia

Lloyd Samson Thole

Assistant Registrar

Patents and Companies Registration Office

P.O. Box 32020

Lusaka, ZAMBIA

E-mail: lloyd@zamtel.zm

ABSTRACT

The Zambian Patents and Companies Registration Office (PACRO) is a semi-autonomous and self-financing executive agency that regulates companies and business registration as well as industrial property. It came to be in its present form in 1998 when it was hived-off from government and commercialized in 1998 under the public service reform. It deals with companies, business names, patents, trademarks and registered industrial designs registrations. Copyrights issues are handled by another office under the Ministry of Information and Broadcasting Services. Most of the patent filings are foreign. On the average, Zambia records about 1200 trademarks, 100 patents and at least 10 registered industrial designs annually. Only 25 per cent of these are local. Zambia, being a Least Developed Country, has a long way to go in making the local stakeholders aware of the importance of intellectual property to national development.

PACRO formulated an IP policy whose work started in 2006 and was adopted in 2010. PACRO is currently reviewing its IP legislation which needs a complete overhaul to bring it to standards to enable Zambia meet the TRIPs obligations by 2013 for general IP requirements and 2016 for pharmaceuticals and essential medicines. An Act has recently been enacted to make PACRO a separate legal entity. IP legislation is weak and outdated making it difficult to enforce IP rights in Zambia. The legislation does not cover protection of geographical indications, service marks, trade secrets, traditional knowledge, traditional and cultural expressions of folklore and genetic resources. Cases of counterfeits and piracy are rife.

Zambia has a great opportunity to make its economy growth through IP development. One of the daunting challenges is the inability to enforce IPRs effectively in the country. It is hoped that the new policy and the revised legislation will help in the administration and enforcement of IPRS in the country.

Keywords: *Policy, Intellectual property, Traditional knowledge, Protocol, Pillaging, Counterfeit, Genetic resources.*

Introduction

As a country, Zambia has not utilized the full potential of Intellectual Property as a tool for social and economic development although intellectual property legislation has been in existence dating back to the pre-independence days. This is so because, despite there being many stakeholders in the Intellectual Property field, a significant number of them are not aware of their role in intellectual property development and utilization. Examples of these are the actors in the informal sector, *viz.*, the Medium, Small-to-Micro scale Enterprises (MSMEs), and in the formal sector, the business community, universities and Research and Development institutions which are the main creators of Intellectual Property. Other stakeholders are in Copyright and Related Rights namely musicians, producers, distribution agents, broadcasters and broadcasting institutions.

In addition, Zambia has an abundance of Traditional Knowledge Genetic Resources, Traditional and Cultural Expressions (Folklore) and whose intellectual property rights can significantly contribute to the social and economic development of the country if properly and effectively harnessed.

The TRIPS Agreement provides that the protection and enforcement of Intellectual Property Rights should contribute to the promotion of technological innovation and to the transfer and dissemination of technology to the mutual advantage of producers and users of technological knowledge in a manner conducive to social and economic welfare, and to a balance of rights and obligations.

The development of any country to a large extent depends on human intellect and what it can create or produce through the use of knowledge as a commodity and capital. Developed nations largely owe their success to a diversity of innovations and inventions, which made mankind progress from the days of the Wright brothers with their test flights to modern day supersonic jets, from the hourglass to deep sea and space chronometers. Probably the best invention for mankind is the wheel. Since its invention, the wheel has been utilized in almost any apparatus one can think of. It has made life easier in more ways than what the inventor could have imagined. Copyright has contributed greatly to the social, economic and cultural development in society in a similar way.

Because of the value which has been recognized in the creation and use of intellectual property for economic development, Zambia has taken strides to develop the Zambian Intellectual Property regime. This has been evidenced by the existing legislation on intellectual property, such as copyrights, patents, trademarks and industrial designs, dating back to the pre-independence era. However, the above legislation has not been effectively applied for successful exploitation of IPRs for national development and economic prosperity. After independence, Zambia joined forces with other countries globally in the creation, acquisition, and protection of intellectual property. Since then, apart from being a member of the World Intellectual Property Organization (WIPO), Zambia has also become a member of the African Regional Intellectual Property Office (ARIPO) based in Harare, Zimbabwe.

Despite a number of national and international developments in the field of Intellectual Property, through the Berne Convention, Harare Protocol, Paris Convention, Rome Convention, and the TRIPs Agreement, Zambia has had no policy to guide the use and exploitation of Intellectual Property. Although the number of patents, trademarks and industrial designs registered under PACRO has increased enormously in the recent past, there has not been a matched growth in intellectual property legislative reform. For instance, the Patents, Trademarks and Registered Industrial Designs Acts have remained unchanged since independence despite the dynamic nature of Intellectual Property.

The Patents and Companies Registration Office

In Zambia, issues of IPR are vested in two Ministries namely the Ministry of Commerce, Trade and Industry and the Ministry of Information and Broadcasting. The Patents, and Companies Registration Office (PACRO) administers the Industrial Property aspect of Intellectual Property Rights, while the Ministry of Information and Broadcasting deals with Copyrights and Neighbouring Rights.

PACRO comprises five (5) units, namely: Accounts; Human Resources and Administration; Information Technology; Industrial Property; and Commercial Unit - the latter being the two (2) core units. The Registrar, who reports to an Advisory Board appointed by the Permanent Secretary responsible for Commerce, Trade and Industry, is the Chief Executive Officer. The Registrar is assisted by two (2) Assistant Registrar's responsible for Commercial (Company and business names registration) and Industrial Property, respectively. During the year under review, Mrs. A.M. Banda-Bobo is the Registrar while the Assistant Registrars are Messrs. Lloyd S. Thole and Mr. Anthony Bwembya. A total of 90 persons are employed by the institution. The institution operates from Mwayi House located along Haile Sellasie Avenue in Longacres, Lusaka.

PACRO was a department within the Ministry, until 1998, when it was transformed into an Executive Agency on similar lines like the UK Patent Office. When it was a department its finances were budgeted and approved by Parliament. Although, the Office generated a lot of revenue nevertheless the funding of the Office was not adequate. Due to financial constraints, the Office could not develop to its full capacity in terms of human resources, equipment and Office space. This resulted in having operational difficulties in the running of the Office. It was almost impossible for the Office to finance the attendance of the Governing Bodies Meeting of WIPO and ARIPO. This had a negative impact on the development of IP Administration in

Zambia. Zambia could not effectively attend most of important meeting pertaining to the IP matters due to lack of financial resources.

PACRO has offices in Ndola, Livingstone, and Chipata in the Copperbelt, Southern and Eastern Provinces of Zambia respectively. With the opening of these regional offices, it will be easy for the public outside Lusaka to register their companies and file their intellectual property applications. In the same reporting period, PACRO was able to make major strides in the preparation of the Intellectual Property Policy which presented to the minister of Commerce Trade and Industry for adoption.

Transformation of the Patents and Companies Registration Office into an Agency

Since independence, the issue of intellectual property has been handled by the PACRO as a department under the ministry of Commerce Trade and Industry. This has meant that PACRO did not have a separate legal entity with its own personality. It also meant that the bureaucracy that is prevalent in government institutions exerted a lot of drawbacks in the operations of the institution. In order to avert this situation, a proposal was advanced to transform PACRO into an institution which would be semi-government and operate on a self-sustaining basis. This would be an Agency which will be autonomous in its operations and have power to sue and be sued. It will also have the ability to own its own property. Until recently, PACRO was not a separate legal entity leading to it experiencing bottlenecks in its operations as it would always have to seek permission from the permanent secretary to carry out some of its activities such as preparation of the budget, preparation of the strategic plan and making changes to the fee structure amongst other things. The Office will now be called the Patents and Companies Registration Agency (PACRA) as opposed to the Patents and Companies Registration Office (PACRO), the name under which it was called previously and will operate under an independent Board.

Intellectual Property Legislation in Zambia

At the moment, Zambia has two separate institutions discharging functions concerning Intellectual Property, namely, the Patents and Companies Registration Office under the ministry of commerce trade and industry which administers industrial property namely trademarks, patents and industrial designs and the Copyright Office under the ministry of information and broadcasting services which administers the Copyright and Performance Rights Act. The Office is responsible for copyright and related rights *viz.*, literary works Musical works, Dramatic works audiovisual works, Recordings Broadcasts Performances of Artists. PACRO is in charge of the protection of Trademarks, Patents and Industrial designs.

PACRO regulates the registration of companies and businesses in Zambia under the Companies Act Cap 388 and Business Names Act Cap 389 of the Laws of Zambia. PACRO also administers Patents, Trademarks and Registered Designs under the Patents Act Cap 400, the Trademarks Act Cap 401 and Registered Designs Act Cap 402 of the Laws of Zambia. By and large, the legislation both for commercial and intellectual property filings is quite old. Legislation for the latter dates to as far back as 1958, six years before Zambia attained its independence. As a result of this, Zambia has had some problems in administering intellectual property effectively. The companies Act was reviewed in 1994, making it slightly better in its ability to facilitate the administration of commercial registrations in form of companies and business names.

Proposed Amendments of the Industrial Property Legislation

The overall objectives of the proposed amendments to IP laws in order to further the integration of Zambia in the multilateral trading system and increase the contribution of IP in general, and TRIPS in particular, to the achievement of sustainable socio-economic growth and development. In brief the amendments will be done in order to:

1. To bring Zambia's IP legislation and regulation in conformity with the World Trade Organisation's TRIPS Agreement and applicable regional and international IP Treaties and Agreements;
2. To propose the legal framework and draft legislation for the protection of relatively new forms of intellectual property that is not covered by the current legislation including geographical indications, trade secrets, and utility models. traditional knowledge, traditional cultural expressions (folklore) and genetic resources; and
3. To build institutional capacity for IP implementing agencies and enhance Zambia's ability to harness IP for socio-economic development.

The process of amending the laws has already commenced and will be completed by next year.

Management and Protection of Trademarks, Patents and Industrial Designs in Zambia

At the moment, an appreciable amount of activity is taking place in the registration and granting Trademarks, Parents and Registered industrial designs. It is important to mention here that most of registrations are for foreign applications emphasizing the fact that Zambia as a Least Developed Country still has a long way to go making the local people aware the importance of intellectual property to national development.

Over the years, the trend of registrations has been fluctuating. Figures 6.1–6.3 showing records of some of the registrations for the year 2004-5, 2005-6 and 2006-7 respectively.

Protection of IPRs through the International and Regional Industrial Property Regime

Zambia is a member of the African Regional Intellectual Property Organisation (ARIPO) which is an intergovernmental Organisation based in Harare, Zimbabwe, established in 1976 to, *inter alia* to promote the harmonization and development of the intellectual property laws, and matters related thereto, appropriate to the needs of its members and of the region as a whole. Zambia is also a member of the World Intellectual Property Organisation (WIPO). It is party to the Berne Convention, the

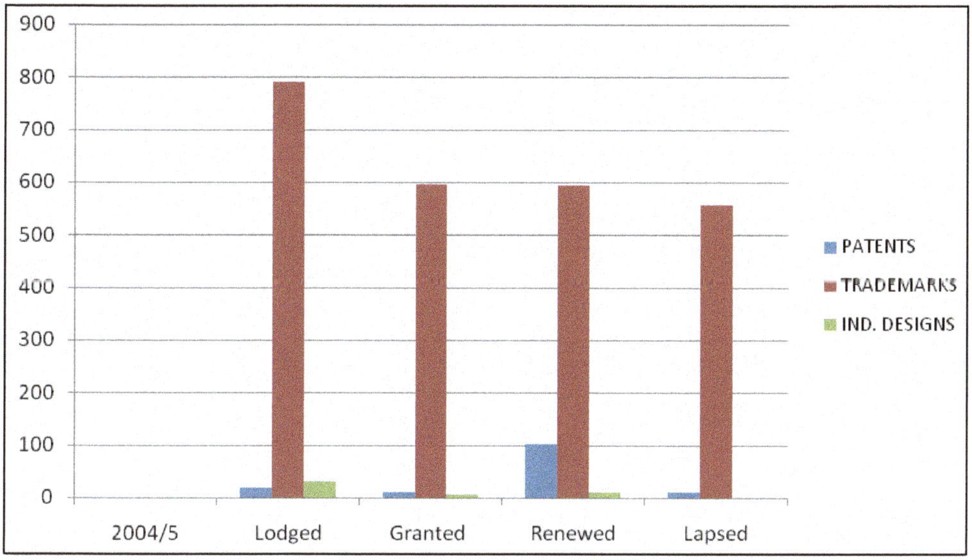

Figure 6.1: IP Statistics for 2004/5

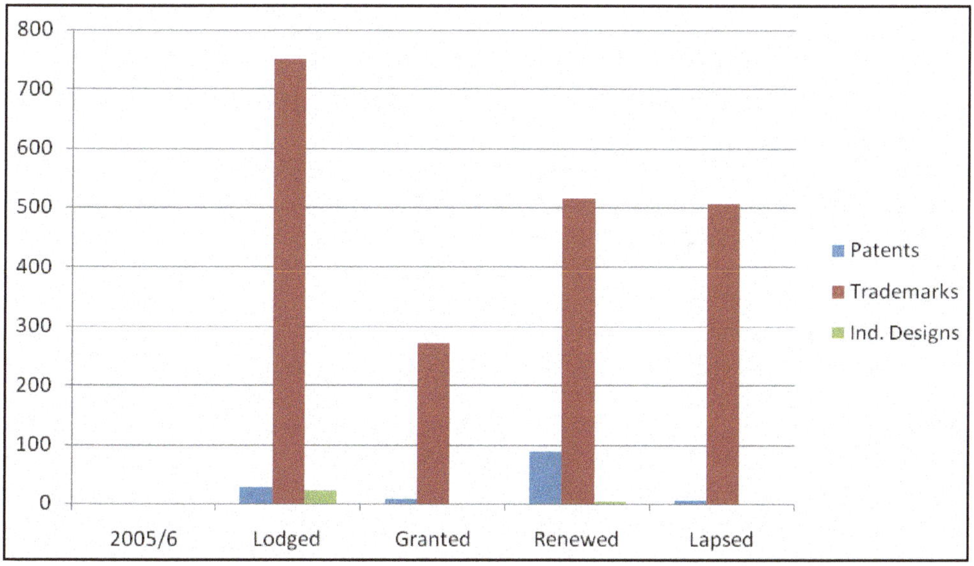

Figure 6.2: IP Statistics for 2005/6

Paris Convention, Patent Cooperation Treaty, Rome Convention, WIPO Producers and Phonograms Treaty and the Madrid Protocol. Zambia handles international applications for trademarks through the Madrid system and Patents through the Patent Cooperation Treaty (PCT). Zambia is not a member of the Hague Convention yet. It is envisaged that Zambia will become a member of UPOV (1978) soon. Zambia also sits on the Coordination Committee of WIPO which is responsible, amongst other things, for staff matters and appointments of officers to the WIPO secretariat.

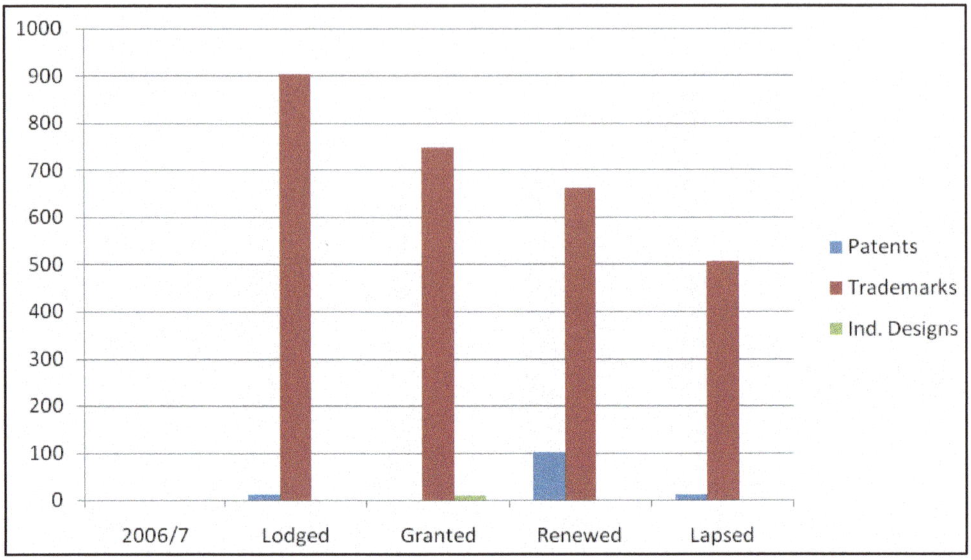

Figure 6.3: IP Statistics for 2006/7

Protection of Genetic Resources, Folklore and Traditional Knowledge

In the recent past, the international community, with WIPO playing a leading role, has come to recognize and realize that traditional knowledge and folklore and genetic resources is a goldmine of plausible intellectual property rights. As the situation stands now, the republic of Zambia is working of the formulation of initiatives to come up with a regime to protect these resources. The protection of Genetic Resources, Traditional Cultural Expressions of folklore and traditional knowledge has eluded the intellectual property system for a long time. The reasons for this are numerous. One of them is that traditional people find it difficult to disclose the details of TK as is required under patent law. Another reason is that TK has never really been documented in the true sense of the word so that users can use it and contribute to its further development. Because of these and other factors, TK has never been well protected leading to its abuse and continued pillaging at the hands of the west. Currently the Zambian Patent office has developed legislation with the help of a consultant to protect these resources. Additionally, the African Regional Intellectual Property Organization (ARIPO) based in Zimbabwe has developed a Protocol to protect these resources. This Protocol will be adopted in a Diplomatic Conference to take place in Namibia in August this year. When this is done, it will be very easy for Zambia to protect its TK resources. Additionally the World Intellectual Property Organization (WIPO) is currently developing an instrument through the Inter-governmental Committee (IGC) on TK, GR and expressions of Folklore. When this is done, it will add further impetus to the drive for the protection of these resources.

The Formulation of the IP Policy for Zambia

Intellectual Property is steadily being recognised as a tool for promoting socio-economic development of nations. Development of nations is gradually becoming dependant on creations of the human mind and the application of such knowledge and ideas as intellectual capital in various areas of development which has significantly contributed to the creation of wealth in these nations. Industrialised countries have effectively utilised Intellectual Property as a tool for social and economic development while developing countries including Zambia have continued to lag behind in this area.

The Ministry of Commerce Trade and Industry (MCTI), through the Patents and Companies Registration Office (PACRO), has for the past four years been working on a National Intellectual Property Policy. The policy was submitted to government for approval late last year and was endorsed in March this year. This policy is aimed at encouraging inventors, innovators and creators to work diligently knowing well that their works will not only give them benefits in form of recognition and royalties but also contribute to national development. The economic development of Zambia will therefore rest on her ability to provide effective legal protection to Intellectual Property and knowledge generated by human intellectual activity and reward inventors, innovators and creators for their efforts.

The Policy will provide guidance on all matters related to Intellectual Property in Zambia and achieve set objectives which include the need to encourage the use and development of appropriate Intellectual Property protection systems, facilitate reform of local legislation and domestication of relevant international agreements that relate to Intellectual Property. It will also encourage research and development and collaboration between research institutions and industry and provide support and guidance from the point of developing and conceptualizing ideas to production and commercialization of goods and services.

The formulation of the Policy was necessitated by the observation that over the years, there has been no Policy in Zambia leading to a major constraint in the appreciation of the importance of the role Intellectual Property plays in the sustainable socio-economic, cultural and industrial development of Zambia. Once functional, the policy will facilitate the revision of Intellectual Property laws in order to make them relevant to issues of protection and management of Intellectual Property benefits, provide a favourable investment climate for both local and foreign companies, encourage innovation and creativity for purposes of boosting social, cultural, industrial and economic development, ensure adequate and appropriate protection, recognition and respect and enforcement of IPRs. This policy will also facilitate the formulation of institutional Intellectual Property policies and provide guidelines for their implementation.

Key Deficiencies in Intellectual Property Management

Human Resources in Intellectual Property

- ☆ The critical mass of intellectual property experts in the country is inadequate.

- PACRO as a registration office only has about eight members of staff who are working in the IP Unit.
- The Copyright Office at the ministry of information and broadcasting services has only two members of staff in the unit.
- As regards the Zambia Police IP Unit, although it has at least 14 members of staff, only one i.e. the head of the unit is adequately trained in the field of IP. The others have recently been enrolled to undertake a certificate course courtesy of the World Intellectual Property Organization which offers an on-line training course in IP.
- Other experts in the intellectual property field are found in legal firms and institutions of higher learning such as the University of Zambia and these are mostly lecturers.

Funding for IP Activities

- Funding for IP activities is very low nationally.
- Activities of awareness are inadequately funded and covered.
- In institutions such as universities and colleges there is a limited amount and types of IP books in the libraries due to inadequate funding.

Limited Awareness in IP

- There is very little awareness on IP issues in the country.
- Most of the members of the public have very little understanding of IP though IP is part of their day to day activities in life.
- IP is very difficult to enforce IPRs in the country as most people do not understand their IP rights.

Training

- Training in IP has from time to time been undertaken by WIPO and ARIPO for people involved in IP work. However, this has not been adequate.
- Additionally the training has only mainly been available abroad and many institutions or individuals cannot afford to pay for these courses.
- The University of Zambia offers some courses in IP at degree level but this is only in the school of law.
- Despite this, the courses in the school of law are optional and many students do not take them as they do not realize their importance until they are out of the university.
- College syllabi have no courses on IP at all.

Lack of Proper Coordination

- There is a fragmentation of laws and institutions carrying out IP work in the country and there is no proper coordination amongst them.

- ☆ As each institutional has a separate piece of legislation administering it, there is often a duplication or omission of work in the administration of IP.
- ☆ The legislation on intellectual property is outdated making it difficult for the country to tackle issues of IP effectively.
- ☆ Universities and other institutions dealing with IP do not have institutional policies to guide them on IP matters.

Way Forward

- ☆ There should be a deliberate policy to inject more funds in the creation of a critical mass of Human Resources trained in Intellectual property.
- ☆ Universities and Colleges syllabi should incorporate IP issues
- ☆ The legislation in IP should, as a matter of urgency, be reviewed and amended or re-enacted if the country has to benefit from IP in the modern world.
- ☆ The government should consider creating or transforming one institution in the country which will act as an apex institution to coordinate activities of IP in the country
- ☆ As there is limited awareness on IP in the country, there should be deliberate efforts to come up with many programmes on radio, television, workshops and seminars to increase awareness on IP in the country.
- ☆ Institutions especially Research and Development institutions should come up with institutional policies to guide issues on IP in line with the National IP Policy.
- ☆ There should also be efforts to harmonize legislation in the country so that IP issues can be addressed in a uniform manner by all concerned.

Conclusion

Many stakeholders are not very familiar with IP issues in the country. However, various awareness initiatives are being undertaken using different fora. When new laws are enacted after amendments, PACRO with its cooperating partners will engage in vigorous provincial awareness campaigns. It is hoped that the new IP laws will be harmonized with other laws which have a bearing on the administration of IP in the country in addition to this, the Patents Office will undertake to engage on a vigorous programme of training for staff to bring them to the levels that will enable them to administer IP issues more effectively.

Acknowledgements

Mrs Anessie Banda-Bobo – Registrar, PACRO

Mr Anthony Bwembya – Assistant Registrar, PACRO

Mr George M. Kanja, Lecturer, University of Lusaka, Zambia

Mrs Chanda – Tembo, Lecturer, University of Zambia, Lusaka

Mr K. Musamvu, Registrar of Copyright, Ministry of Information and Broadcasting Services, Lusaka, Zambia.

References

The National Intellectual Policy of Zambia, 2010, PACRO

Intellectual Property Law, George M. Kanja, 2006, University of Zambia

PACRO Annual Reports, 2006-2008.

Abbreviations

ARIPO: African Regional Intellectual Property Organization

IGC: Inter-governmental Committee on Traditional Knowledge, Genetic Resources and Expressions of Folklore

IP: Intellectual Property

IPRs: Intellectual Property Rights

MCTI: Ministry of Commerce Trade and Industry

MIBS: Ministry of Information and Broadcasting Services

MSMEs: Medium, Small-to-Micro scale Enterprises

PACRA: Patents and Companies Registration Agency

PACRO: Patents and Companies Registration Office

PCT: Patent Cooperation Treaty

TK: Traditional Knowledge

TRIPs: Trade-Related Aspects of Intellectual Property

UPOV: Union for the Protection of Plant Varieties

WIPO: World Intellectual Property Organisation

Chapter 7

IPR Management in Tanzania: The Role of Tanzania Intellectual Property Advisory Services and Information Centre (TIPASIC)

Judith F. Kadege

Research Officer (IP)

Tanzania Commission for Science and Technology (COSTECH)

P.O. Box 4302, Dar-es-Salaam, Tanzania

E-mail: jkadege@costech.or.tz

ABSTRACT

In Tanzania, Intellectual Property (IP) is a non-union matter. It is also a new notion among the majority of the persons within the country, despite the fact that the concept of IP legal system has been there since the times of colonial rule. During that era, various Laws[1] concerning intellectual property rights were introduced and implemented in the then called Tanganyika.[2]

This chapter discusses and examines issues of intellectual property regime in Tanzania (Mainland) including IPR administration and legislation of Tanzania. In Tanzania, lack of awareness and knowledge of IP/IPR matters among the majority is still the major problem despite the little rise of the basics and knowledge among the academics, legal, scientists, artists etc. In addition, the paper looks at how IP is

1 Naming a few: - The Copyright Legislation Cap. 218 of 1924; Trade Marks Legislation of 1922 and the Patent Legislation Cap. 217 of 1924 etc

2 After the union between Tanganyika and Revolutionary government of Zanzibar, the country **Tanzania** was born.

administered and how the responsible offices are mandated and function in the administration of IP in Tanzania (Mainland).

Further, this chapter considers the current initiatives done in Tanzania, by few selected institutions, to promote and utilize IP in various sectors such as agricultural, technological, artistic etc. these efforts have helped the stakeholders within this particular field to have at least awareness on the said area i.e. Intellectual Property, however more is needed to be done for the betterment and sustainable economic development of the country. Among these plans, is the establishment of Tanzania Intellectual Property Advisory Services and Information Centre (TIPASIC) which was established in 2007 with the main aim of providing services and help on all matters concerning Intellectual Property (IP) while putting emphasis on accessing information contained in Patent documents for technological and economic advancements in Tanzania.

Keywords: *Intellectual Property, IPR administration and Legislation in Tanzania, Role of TIPASIC, IP awareness, poverty reduction, Tanzania.*

Introduction

Historical Background of the United Republic of Tanzania

The United Republic of Tanzania is located in Eastern Africa between longitude 29 degrees and 41 degrees East, latitude 1 degree and 12 degrees South. The United Republic of Tanzania comprises two sovereign states namely, Tanganyika and Zanzibar. The two entities agreed voluntarily to unite and formed the United of Tanzania on 26 April, 1964. The United Republic of Tanzania has a population of approximately 40 million people and 51 per cent of them are females. The Mainland has a population of 39m and Zanzibar has 1m people. The United Republic of Tanzania stretches over 364,900 square miles of land and has massive stretches of inland lake waters including Lakes Victoria, Tanganyika, Nyasa and many others. While the then Tanganyika won her independent on 9th December, 1961 from the British colonial rule, Zanzibar became independent in 1963.

The key economic sectors which contribute significantly to the country's Gross Domestic Product include Agriculture, Mining, Energy, Infrastructure and Tourism. Agriculture contributes 50 per cent to the national income, produces 75 per cent of the country's merchandise exports and employs 80 per cent of the rural populace. Agriculture is one of the leading sectors which have linkages with almost all the remaining sectors through forward linkages namely, agro-processing, provision of consumption and exports, provision of food and raw materials as inputs to industries and markets to manufactures farm implements.

IP Legal Systems and Administration in Tanzania

Basically the intellectual property law aims at safeguarding creators and other stakeholders of intellectual goods and services by granting them time-limited rights to control the use made of the intellectual creation productions. Hence, those rights are not applicable to physical objects in which the creation may be embodied into.

There are generally two categories of Intellectual Property, Industrial Property and Copyright. Whereas the Industrial Property deals with the protection of trademarks, industrial designs, and geographic indications of source, inventions (patents), Copyright deals with protection of literary and artistic works such as novels, poems and plays, films, musical works and artistic works which include drawings, paintings, photographs and sculptures, and architectural designs.

Under copyright, there are related rights which deals with those rights of performing artists in their performances, producers, and those of broadcasters in radio and television programs.

IP Legal Systems in Tanzania

The attainment of independence of Tanganyika in 1961 did not reflect into major changes in the context of IP legislations. To date, legislative framework governing IP in both Tanzania (Mainland) and Zanzibar comprise of the following:

A. Tanzania (Mainland)

1. The Trade Marks and Service Marks Act 1986 [R.E 2000]
2. The Patent Act 1987 Cap. 217 [R. E 2000]
3. The Copyright and Neighbouring Right Act No. 7 (1999)
4. Plant Breeder's Rights Act 2002
5. The Fair Competition Act, 2003 (Act No. 8 of 2003)
6. The Merchandise Act 2007

B. Tanzania (Zanzibar)

1. United Kingdom Designs (Protection) Decree, cap. 158 (27 of 1937)
2. Trade Marks Decree, cap. 159 of 1956
3. The Patent Decree Cap. 157 of 1958
4. The Zanzibar Industrial Property Act 2008

Tanzania is one of the signatories to various Treaties and has Memberships with respective organizations such as:

1. Paris Convention (Industrial Property), since June 1963.
2. ARIPO: Member of ARIPO, since October 1983.
3. WIPO Convention, since December 1983
4. Berne Convention (Literary and Artistic Works), since July 1994.
5. WTO: Member and Signatory to TRIPS Agreement, since January 1995
6. Nice Agreement (International Classification of Goods and Services), since 1999
7. PCT (Patents), since September, 1999.

Institutional Structure for IP Administration in Tanzania

In 1980s, Tanzania government committed itself to economic reforms when the economy was transformed from a centrally planned economy to a market economy. With the introduction of appropriate market economy policies, such as liberalization and privatization as means to sustain economic development of the country where competition replaced monopolistic practices in the markets for both goods and services and consequently the Government, in the late 1990s, began intensive efforts to create a practicable regulatory framework in the country thereof.

In Tanzania, there are various institutions that deal with matters relating to Intellectual Property Rights, including government agencies and private sectors. However, it's only Government's agencies either under Ministries concerned or mandate by the government independently, that have been given powers to deal with the administration of IP such as Copyright, Patent and Breeders' Rights etc.

The Copyright Society of Tanzania (COSOTA)

The legislature of the United Republic of Tanzania enacted the Copyright and Neighbouring Act, No. 7 of 1999, in April 1999. The President gave his assent in June 1999 and the law became operational from December 31, 1999.

Copyright Society of Tanzania (COSOTA) was established under Section 46 of the Copyright and Neighboring Rights Act, 1999 (No. 7 of 1999 RE 218 of 2002) and came into operation on 24th October, 2001. COSOTA has been vested with power to administer Copyright under the Ministry of Industry, Trade and Marketing.

The main functions of COSOTA include:

1. Promoting and protecting authors, performers, publishers, translator of works, and broadcasters.
2. Determining the minimum rates of royalties to be charged on uses of works and performances of its members.
3. Collecting and distributing of royalties in respect of uses of the works of its members.
4. Printing, publicizing and circulating of information and the rights of its members.
5. Searching, identifying and publicizing rights of owners, and defending them.
6. Maintaining registers of works, productions and associations of its members.

COSOTA is a multidisciplinary society operating as a Copyright office of the United Republic of Tanzania a well as a collective Management organization. As a Copyright office, COSOTA is tasked with activities such as the elimination of Copyright Piracy- infringement of Copyright in Musical, Cinematographic, Literary and Artistic Works whereas as a Collective Management activities include registration of members and their works; and Collection of royalties from users of Musical and Cinematographic works in commercial establishments such as discotheques, bars, entertainment halls, hotels, broadcasting stations etc.

COSOTA also has a mandate under the Act[3] to protect both local and foreign firms as well as agreements of reciprocal representation with foreign sister societies.

The law on copyright, neighbouring rights and expressions of folklore complies with the Berne convention (1886) and the TRIPS Agreement of the World Trade Organization (1994). The law aims at:

1 Protecting of the moral and economical interests of authors (creators) relating to their works.
2. Providing protection for expression of folklore.
3. Protecting interests of performing artists, producers of cassettes and broadcasting organizations.
4. Providing for civil remedies and criminal sanctions against infringers and pirates.

Business Registrations and Licensing Agency (BRELA)

Competition usually leads to product and systems innovations for better quality products and services in the market and also value for money, all for the benefit of consumers. Nonetheless regulation and facilitation are essential factors for a vibrant, sustainable and effective market economy, to ensure participants in the market economy can compete fairly.

In realizing this, the Tanzania Ggovernment decided to establish Government Executive Agencies to administer IP in the country. These agencies include: The Business Registrations and Licensing Agency (BRELA), The Fair Competition Commission (FCC), Plant Breeders' Rights Registry (PBRR) etc.

The Business Registrations and Licensing Agency (BRELA) is a Government Executive Agency established under the Government Executive Agencies Act No. 30 of 1997. It was established on the 28th of October 1999 by Government Notice No. 294 and it was officially inaugurated in December 1999. The Agency is under the Ministry of Industry, Trade and Marketing has been given a power to administer industrial property matters such as Patents, Trade Marks and Designs etc. BRELA has a division that deals with all matters related to Intellectual Property (industrial property).

Others functions of BRELA include:

1. Administration of companies and business names laws.
2. Stimulation of scientific and technological inventiveness and innovation and encourage technology transfer.
3. Administration of Intellectual Property Laws.
4. Regulation of business by administering business and industrial licensing laws.
5. Facilitation of local and foreign business investment.

---

3 Section 3, 9, 47, and 48 of the Copyright and Neighboring Rights Act, 1999 (No. 7 of 1999 RE 218 of 2002).

The Fair Competition Commission (FCC)

The Fair Competition Commission is an independent government body, under the Ministry of Industry, Trade and Marketing, established by Government of the United Republic of Tanzania through the Fair Competition Act, 2003 (Act No. 8 of 2003) to promote and protect effective competition in trade and commerce and to protect consumers against unfair and misleading market conduct throughout Mainland Tanzania subject to the usual generic competition enforcing requirements of independence, accountability, transparency and due process. The ultimate goal of the Act is to increase efficiency in the production, distribution and supply of goods and services. This Act came into operation in May 12, 2004.

The establishment of FCC has been seen as a significant step in Tanzania's effort to establish a market economy. Functions of the Commission include:

1. Administer the Fair Competition Act.
2. develop and promote policies for enhancing competition and consumer welfare.
3. Promote public knowledge, awareness and understanding of the obligations, rights and remedies under the Act and the duties, functions and activities of the Commission.
4. Carry out inquiries studies and research into matters relating to competition and the Protection of the interests of consumers.
5. Study government policies, procedures and programs, legislations and proposals for legislation so as to assess their effects on competition and consumer welfare and publicize the results of such studies.
6. Investigate impediments to competition, including entry into and exit from markets, in the economy as a whole or in particular sectors and publicize the results of such investigations.
7. Investigate policies, procedures and programs of regulatory authorities so as to assess their effects on competition and consumer welfare and publicize the results of such studies.
8. Participate in deliberations and proceedings of the government, government commissions, regulatory authorities and other bodies in relation to competition and consumer welfare.
9. Control, manage and efficiently perform the functions of the Commission under the Act.

Furthermore, the FCC deals with all issues of anti-competitive conduct, abuse of dominance and has provision for curtailing mergers and acquisitions if outcome is likely to create dominance in the market or lead to uncompetitive behavior.

Nonetheless, the fundamental aim of these institutions is to ensure businesses operate in accordance with the laid down laws, regulations and sound commercial principles. Thus, through working hand in hand with other mandated principal stakeholders such as Tanzania Revenue Authority (TRA), Tanzania Bureau of Standards (TBS), Tanzania Food and Drugs Authority (TFDA), Attorney General's

Chamber and the Police Force, campaigns and essential actions to stop counterfeit goods in the market (country) have been successful despite the overwhelming challenges.

Plant Breeders' Rights Registry (PBRR)

The main legislation for protection of plant breeders' rights in Tanzania is the Protection of New Plant Varieties (Plant Breeders' Rights) Act, 2002. This Act introduced new and independent protection system for plant varieties not registered under the Patents Act.

The Plant Breeders' Rights Act of 2002 is a *sui generis* piece of legislation that provides for a voluntary system intended to enhance plant breeding activity in the public and private sectors. The Act adopts, to a large extent, major provisions of the UPOV Convention of 1991 although some differences are notable. It was officially declared operation on 1st February, 2004.

Main objectives of the Act include, but not limited to:

1. Establishment of a registry of plant breeders' rights.
2. Promotion of plant breeding and facilitation of agricultural advancements through the grant and regulation of plant breeders' rights and for matters connected herewith.

Section 5 of the PBRR Act establishes an independent plant variety protection office known as the Plant Breeders' Rights Registry, under the Ministry of Agriculture, Food Security and Cooperatives, which is responsible for administration of plant variety protection system. This Office falls under this Ministry due to the fact that it (Ministry) is better placed to deal with plant variety matters than any other government Ministry.

The functions of this Registry consist of:

1. Granting plant breeders' rights.
2. Maintaining a register and provide information on plant breeders' rights issued in Tanzania.
3. Facilitating transfer and licensing of plant breeders' rights.
4. Collaborating with local and international bodies whose functions relate to plant breeders rights matters.
5. Establishing a documentation centre for the purposes of dissemination of information on plant breeders rights.
6. Performing such other functions as are necessary for the furtherance the objectives of the Act, etc.

Other Agencies and Institutions Involved in IP Matters Include

1. Judiciary (High Court, Commercial Court, District Courts - Copyright).
2. Police - Apprehension and prosecution.
3. Customs - imports inspection.

4. Tanzania Bureau of Standards (TBS) - Consumer protection and identification marks.
5. Commission for Science and Technology (COSTECH)- Technology Transfer and Inventions.
6. Other Various Research and Development institutions.

The Tanzania Intellectual Property Advisory Services and Information Centre (TIPASIC)

Establishment and Background of TIPASIC

After realizing the importance of patent information in technological advances and business opportunities, the Business Registrations and Licensing Agency (BRELA) and the Tanzania Commission for Science and Technology (COSTECH) in collaboration with the World Intellectual Property Organization (WIPO) initiated a move to establish Intellectual Property (IP) advisory services in Tanzania. The Idea started by conducting a two - day workshop in October 2005 which focused on Patent Documentation: A contribution for business activities and effective research and development.

After the workshop, WIPO sent a mission to Tanzania, which visited COSTECH and the University of Dar es Salaam. After consultation the mission proposed COSTECH to host the centre (TIPASIC). COSTECH was requested to provide an office to accommodate the centre. In July 2006 WIPO sent some equipment for the centre, the centre is now equipped with computer, saver and other communication facilities.

As a policy for capacity building and sensitization, WIPO conducted a training workshop on IP Advisory Services and Information Centre Skills Development Training in Dar-es-salaam, in December 2006 and also WIPO provided training, in Geneva, for two staff responsible for running the Centre in March 2007. The Centre was officially established in March 27, 2007 at COSTECH.

Role of TIPASIC

Since its establishment, TIPASIC has been in a front line in providing services and advice regarding IP in particular on Patent information. However, the Centre has also assisted in providing information on IPRs such as Copyright, Trademarks etc and creating awareness to various clients and stakeholders on issues relating to technology transfer and innovation. Other roles of TIPASIC include:

Creating Awareness and Sensitization on IP Issues

The Centre has organized training, seminars and workshops, and still do the same, on issues of intellectual property rights and technology transfer. Also, the Centre encourages the establishment of technology transfer offices in various institutions particularly in R and D institutions.

Providing Advice to Clients on Preparation of Patent Application Documents

One of the major functions of TIPASIC is to assist clients in preparation of patent documents and application of patents.

Providing Patent Information

The Centre also assist clients to search for patent information related to their inventions, innovations or research works from on-line databases and CD-ROMs. Facilitate walk-in information search through internet. Printing copies at cited documents (at a cost).

Technology Transfer: Selection and Agreements

Clarifying technology transfer agreement clauses and giving advice on selection of appropriate technology.

Advisory Services on IP and International Conventions

The Clients also benefit the advice on IP in general and also legal advice on IP laws of Tanzania and IP related International Conventions and Organizations that Tanzania is a member. This is normally done by giving introductory advice and information on both the IP legal framework of Tanzania and WIPO (including IP related Conventions) without forgetting the operation and applicability of these laws. This has helped various clients, innovators, creators etc simply because it has been a source of knowledge and information on IP legislation of Tanzania and other international Conventions on IP, to many of them.

Since TIPASIC has been in operation for three years, from 2007, during these years the centre has been able to organize indoor seminars and workshops, the main aim being that of creating awareness to the society (public) however more emphasis need be focused on R &D institutions as it is through R and D that the society is easily educated and also the dissemination of information is much better vide that format.

These indoor seminars aim at enriching the society with the knowledge of Intellectual Property. Reaching the society at their own place makes it convenient for all of them to benefit from the services of TIPASIC. 101 participants have benefited from these seminars in six research and development institutions in Dar-es-Salaam and Kibaha district. Generally speaking, this can be considered to be the major achievement of the Centre simply because by educating and sharing information on concerned areas many participants from various institutions such as R&D institutions, Universities, NGO's, Companies etc have been able to take more steps on concerned areas at the same time making efforts on dissemination of information on IP related matters in their institutions.

TIPASIC's Relationship and Initiative

BRELA - TIPASIC Linkage

TIPASIC has a good linkage with Business Registration and Licensing Agency (BRELA). Due to limited staff and expertise, majority of client approaching BRELA for patent applications are directed to TIPASIC for advice and technical support on how to proceed with patent applications before they lodge their application to BRELA.

Additionally, TIPASIC has started to compile all patented documents registered at BRELA since the establishment of the Patent Act in 1987 with intention of establishing a database of technologies. So far, there are numerous practical case studies in patent application support done by TIPASIC to innovators (clients) that use the centre's advice and information services on IPR. Few examples include:

Table 7.1: Case studies in Patent Support

Sl.No.	*Submitted Claim*	*IPR Status*
1.	A device for producing heat energy from biomaterial and fuel wood	On the process to be Patented
2.	A device for generation of electricity	On the process to be Patented
3.	An improved method of using a rechargeable battery for domestic electricity needs	On the process to be Patented

R&D Institutions – TIPASIC Initiatives

As a strategy of encouraging and supporting establishment of institutional IP policies and Intellectual Property Management Offices (IPMO) at R&D institutions, TIPASIC has started initiatives of establishing focal points within R&D Institutions. These focal points will have two staffs fairly knowledgeable on IPR issues. The initiative started with four R&D institutions.

The agreement has been made between these two parties such that TIPASIC will provide training and necessary information (Books, CDs etc) and Institutions will allocate time to one or two staff and at least one computer specifically committed for IPR issues within the institution.

TIPASIC will use focal personnel to facilitate establishment of Institutions IP policy and IPMO.

Incorporating IPR in COSTECH Operations

The Centre for Development and Transfer of Technology (CDTT) is among the four departments of COSTECH. The centre has been using multiple approaches in promoting innovation and technology development including awarding credible innovations, supporting prototype development and providing necessary workshop tools and equipments to group of innovators. However, before issuing any support to an innovator, the centre commission experts to evaluate and verify the claims. Since establishment of the TIPASIC all innovators are exposed to IPR issues and are advised to the possible options to proceed.

In addition, all external experts are supposed to sign non-disclosure agreements before starting the evaluation of any new claims submitted at COSTECH.

Challenges of IP in Tanzania

There are numerous challenges that face both TIPASIC and Tanzania as a country, in not only accomplishing the promotion of utilization of patent information within the country but also in awareness creation of IPRs matters, this includes but not limited to:

Low Intellectual Property Awareness Level

There is lack of public awareness on the importance of IP and complying with the Intellectual Property laws. Though the idea of Intellectual Property has been there long ago since the colonial era (particularly, the laws on IP) it is still a new concept in the country as practically the majority do not have the basic knowledge of IPRs. For example: there is a critical problem of piracy though the people are yet to realize the magnitude of this problem; in fact they do not see that copying ones' work is wrong. Inadequate understanding of IP language by both TIPASIC's clients as well as the citizenry in general leads to spending a lot of time to provide service to the clientele as this is a prerequisite for one to appreciate the service rendered on IPR.

However, due to these efforts/work done (educating) by centers and government's office that deal with administration of IP and information on related matters the awareness and basic knowledge on IPR issues *i.e.*, protection, rights, infringement, sanction (punishment) etc, has at least improved despite only a few have this knowledge.

Outdated IP Laws

Tanzania is faced with old-fashioned laws on IP. It is commonly known that the effectiveness of laws depend on the level of their enforcement. Tanzania's intellectual property laws are good reference tools and potential instruments for protecting people's rights. The relevant intellectual property laws are in the books but due to various shortcomings including institutional weaknesses, the outdated laws are not fully and systematically enforced. Nonetheless, the government has started efforts in reviewing some of these laws including the Copyright and Neighbouring Rights Act of 1999 etc. the push for the review of these laws has been the efforts done by the creators and stakeholders themselves as a result of awareness programs obtained from offices and centers that deal with IP issues.

Other challenges include:

- ✰ Limited Capacity particularly in the National Offices to undertake and adopt Policies developed n the Region.
- ✰ Wrong views of IP (IP is too burdensome, IP is for the Rich, IP is for Lawyers etc.).
- ✰ Lack of a National IP Policy.
- ✰ Inadequate experts in IP issues.
- ✰ Lack of IP management offices in R&D institutions and industries.
- ✰ Lack of fast internet connectivity at R&D institutions and sometimes lack of internet connectivity at all. This discourages users/clients in proceeding to do patent information search.

Recommendation and Conclusions

Despite the legal framework and the necessary institutional framework for IPR administration in Tanzania is in operation, other IP implementing agents such as the police department, the customs office and the judiciary need adequate preparation in order to be able to control infringement of intellectual property rights. Also, Tanzania has not yet taken upon the Patent system and the TRIPS Agreement research exemption as a result researchers in universities and R&D institutions do not make effective use of the system and the exemption. The utilization of the patent system especially its information function could make the Tanzanian-researchers conduct research work solve the problems speed up economic growth with the little resources available.

Intellectual Property covers a wide area and is a cross-cutting issue with huge importance economically, culturally and socially. The challenges faced by Tanzania, which are also common in many developing countries, highlight the value and need for further actions in the IPR administration and management in respective countries, Tanzania being among them. The need for these actions is high and thus the following are the concerned recommendations regarding this matter, in Tanzania:

- There is a need of a comprehensive National intellectual property policy.
- IP should be included in the curricula from simple definition at the primary level education to high education levels of creating, protection and exploitation of intellectual property.
- Creation of more IP awareness for the public and among the research community be encouraged.
- Establishment of intellectual property management offices in universities and R&D institutions and other tertiary learning institutions.
- Government's should commit itself on technical, physical infrastructure development and financial support for institutional capacity building in IPR.
- Training of more personnel in the area of IPR. Government should devote to human resources development/capacity building through provision of short and long term training sessions.

The process of managing property rights in the country (Tanzania) will be successful only if all the key players *i.e.* The Government, Public, Stakeholders and IP owners (artists) will have respect for intellectual properties. The centers and offices that administer IP, the legal framework on IP must reflect the changes in the current IP world and hence must be up to date so as to relate with the needs. Also, in universities and research organizations, an obvious understanding of the roles of each stakeholder inside the institutions, as well as outside, is important. Along with these, continuous communication must exist with the external stakeholders including the government, the industry, the legal and the financial institutions. Partnership must be forged between the different groups so that the atmosphere of creativity prospers.

Additionally, not only awareness of IPR is needed and/or of importance in today's countries like Tanzania but also efforts need to be done to create awareness of the existence of IPR documents, rights under the IP laws, benefits etc., for example the patent documentation, among the scientists, technologist, inventors, innovators and the general public. Likewise, most of the areas of IP encourage creativity and innovations, with these efforts and plans it is believed that not only knowledge but also creativity, access of information being available and used properly will lead to the lessening or completely removal of poverty in Tanzania and other developing countries with similar obstacles.

Acknowledgements

I would like to acknowledge the contributions of Tanzania Intellectual Property Advisory Services and Information Centre (TIPASIC), Centre for Development and Technology Transfer (CDTT) and my employer Tanzania Commission for Science and Technology (COSTECH) for facilitating and providing me with necessary and important information and links in the area of IP management and administration in Tanzania, which has added value to the report I am submitting at this forum. I would also like to register my appreciation to NAM S&T Centre for giving me this opportunity to participate in this capacity building programme which will go a long way to establish a strong base for my understanding and appreciating the importance of IP in socio-economic development.

References

BRELA and COSTECH (2004) "Report on the 5th African Union Intellectual Property and Technology Day and Proceedings of the IPR Workshop." September 12- 13, Dar es Salaam, Tanzania.

Kalanje, C. (2006) "Role of Intellectual property in Innovation and New Product Development" Retrieved 25 October, 2006 (http://www.wipo.int/sme/en/documents/pdf/ip_innovation_ development.pdf).

NBS and ORC Macro. 2005. *Tanzania Demographic and Health Survey, 2004-2005.* National Bureau of Statistics, Dar es Salaam. (www.nbs.or.tz)

Ngwediagi P., Establishment of Plant Breeders' Rights System In Tanzania: Achievements and Challenges In: CAS-IP NPI. 2009. Institutionalization of Intellectual Property Management: Case Studies from four Agricultural Research Institutions in Developing Countries. CAS-IP, Rome, Italy. (http://www.kilimo.go.tz)

Shemdoe, G., The Role of Intellectual Property System in Innovations in Developing Countries. (http://www.innovationafrica.net)

Shemdoe, G., Kasonta, J., and Chuwa, P. (2005) "Report on Current National Policy Frameworks on Intellectual Property (IP) Management and IP Capacity Building Initiatives in Tanzania." *Institutional Management of Intellectual Property in Biotechnology Research& Development in East Africa: Country Case Studies*, BIO-EARN Programme Publication, October 2005.

E-source

http://www.tipasic.or.tz

http://www.costech.or.tz

www.brela-tz.org/htmls/IPJ per cent 205 per cent 20 (part per cent 20two

Abbreviations

ARIPO: African Regional Intellectual Property Organization
BRELA: Business Registrations and Licensing Agency
CDTT: Centre for Development and Technology Transfer
COSOTA: Copyright Society of Tanzania
COSTECH: Commission for Science and Technology
FCC: Fair Competition Commission
IPR: Intellectual Property Rights
IPMO: Intellectual Property Management Offices
PCT: Patent Cooperation Treaty
R&D: Research and Development
NGO's: Non- Government Organizations
TBS: Tanzania Bureau of Standards
TFDA: Tanzania Food and Drugs Authority
TIPASIC: Tanzania Intellectual Property Advisory Services and Information Centre
TRA: Tanzania Revenue Authority
TRIPs : Trade- Related Aspects of Intellectual Property Rights Agreement
WIPO: World Intellectual Property Organization
WTO: World Trade Organization

Chapter 8
Protection of Traditional Knowledge in Uganda

Stephen Rwagweri Atwooki
Engabu Za Tooro,
P.O. Box 886, Fort Portal, Uganda
E-mail: engabuzatooro@gmail.com

ABSTRACT

Uganda has diverse and rich traditional knowledge, covering socio-economic aspects of life like music, folklore, apprenticeship, handcraft, production, medicine, etc.

The modern civilization had threefold effects on the traditional knowledge; Some forms of traditional knowledge steadily disappearing, some remaining static and some prevailing through adaptation and integration in modern knowledge.

The Copyright and Intellectual Property rights laws in Uganda stem from the British Copyrights laws and Intellectual Property regimes. As a result, these laws are based on western modes of authorship and ownership and could not recognize Ugandan traditional forms of knowledge like oral traditions and folklore.

Traditional knowledge and Folklore in Uganda is not protected. The technical explanation is that traditional knowledge and folklore doesn't meet the established international standards and criteria of copyright protection which define what, how and who can be protected and for how long. This has been just an intellectual debate among the lawyers and scholars of copyright law. The traditional knowledge sector itself is not developed to generate organized and sensitized practitioners to participate in the debate and generate demand for protection.

The production, presentation and distribution of different forms of traditional knowledge is not organized and developed to secure themselves for protection.

Actors in traditional knowledge sector like cultural and traditional institutions and civic groups are not sensitized, organized and coordinated enough to pursue a common mission of organizing the sector and securing protection.

Some forms of traditional knowledge increasingly get threats of extinction due to failure to be relevant in the contemporary life, failure to cope with effects of high-tech modern communication, emerging hostile beliefs and value systems and imported forms of entertainment.

As traditional knowledge sector attempt to pursue protection, challenging questions emerge. Such questions include the definition of public domain, how rights are shared between original author and the one who re-invests and makes folklore relevant today? How to reconcile need for originality and need for adaptation?

It is recommended that Uganda first focus on programmes to re-invent and organize traditional knowledge so that it can meet some standards for protection and a study be undertaken of possibility of evolving a unique regime for protecting traditional knowledge.

Programmes of research, documentation and digital data systems for Uganda's traditional knowledge should be developed like it has been done in India and South Africa.

Keywords: *Uganda, Patent, Law, Knowledge.*

Introduction

Uganda

Uganda is a landlocked East African country with a land area of 241,039 square kilometers and a population of 30 million people. The country's GDP per capita stands at US $ 514.

Engabu Za Tooro

Engabu Za Tooro is a Ugandan non-governmental organization promoting research and documentation of traditional knowledge and folklore. It is also accredited to WIPO's Inter-governmental Committee on Intellectual Property and Genetic Resources, Traditional Knowledge and Folklore and UNESCO to offer advisory services to the Inter-governmental Committee on Intangible Cultural Heritage.

The organization is currently establishing a cultural centre to be an epicenter of traditional knowledge research and documentation and digital database development.

Why Concern with Traditional Knowledge

Majority of indigenous and grassroots population in Uganda still rely on traditional knowledge for information and livelihood. Traditional knowledge informs scientific innovation. Therefore, there is need to give traditional knowledge its appropriate recognition.

The State of Traditional Knowledge in Uganda

Like any African society, Uganda has a rich and diverse stock of traditional knowledge in all human sectors like medicine, music, production, folklore, dance, apprenticeship and craftsmanship.

Uganda also has traditional and cultural institutions mandated to safeguard and perpetuate the traditional knowledge. Historically, Uganda had five major Kingdoms that included Bunyoro, Buganda, Ankole, Tooro and Busoga and chiefdoms in several parts of the country. These institutions traditionally had political and administrative powers which led to collision with colonial and post-colonial central governments leading to their abolition in 1967. There was a three decade gap and then they were re-instated in 1993 as symbols of history and mandated to promote people's cultural heritage.

However, most of these institutions to date have not moved beyond superficial mobilization to visible programmes that incarnate and build on people's traditional knowledge. They are more often constrained by lack of funding and lured by political expediency.

We note that despite weaknesses and limitations, there are some emerging programmes in some kingdoms that build on, and promote traditional knowledge; examples are the *Burungi Bwansi, Senga* and *Ekisakate* programmes by Buganda Kingdom (COMPASS 2009).

In Uganda there are some traditionalists and development workers who have organized themselves in regional and national associations to safeguard and promote some form of traditional knowledge.

These include;

- ☆ The traditional herbalists who organized themselves into Buganda *Nedaggala lyaayo.*
- ☆ Performers of traditional dances and folklore who organized into Uganda Development Theatre Association (UDTA).
- ☆ The community and private museum operators who have organized themselves into the Uganda Community Museum Association (UCOMA)
- ☆ Promoters of indigenous languages who have organized as Society for Advancement of Ugandan Languages (SAUL).

According to the mapping exercise done in 2007 by Cross Cultural Foundation of Uganda (CCFU), 23 organizations and projects in the country were using indigenous or traditional knowledge in development interventions in different socio-economic areas. One example of such project is Engabu Za Tooro in Western Uganda which creatively uses an indigenous tradition of an ancient Tooro heroine – Koogere – to promote women participation in development (CCFU, 2008).

CCFU went ahead to network such initiatives for information exchange and capacity building. However, these efforts need to be consolidated and coordinated by networks to build a shared vision and a common voice.

The Threat of Extinction of Forms of Traditional Knowledge

Some aspects of traditional knowledge and folklore are facing threats of extinction. Many unique forms of traditional knowledge of minority communities with rich cultural diversity are endangered. Such forms include folklore and indigenous languages.

The threat, in the first place, is caused by the failure of some forms of traditional knowledge to adapt to the changing environment. This is actually the failure of the owners of the traditional knowledge to re-invent it so that it remains relevant in the changing circumstances. (Duncan Smallman Edinburg) quoted in (BBC 2009 – Saving Native American Languages) argues that "a language survives only when it changes with the times." Of course language is the custody of people's traditional knowledge.

The second source of threat is the sophiscated modern communication technology. How do oral forms survive and remain valid in a high tech world in which communication is effected through email and cell phones?

The third source of threat comes from the mighty of imported forms of entertainment. Uganda like many other developing countries is often a dumping ground for a lot of undesirable forms of entertainment mainly from the West. These are packaged to be attractive especially to the youth. They eventually dominate the industry and the space, displacing traditional expression which is often not repackaged to be equally attractive.

The fourth source of threat is emerging beliefs and values that condemn some forms of traditional knowledge one by one. Many of these beliefs originate from, or are influenced by the colonial euro-centric mentality which used to blindly condemn anything indigenous and African as primitive and undesirable. So in most cases the judgment and condemnation of some forms of traditional knowledge, by the adherents of such beliefs and values, is subjective and based on prejudices.

In western Uganda, there is a religious sect which emerged only a few years back and it has condemned the local tradition and practice of pet names. It considers such, as satanic and calls the followers to condemn them. The pet name tradition is very rich and unique traditional heritage that is known to be found among only three communities in the whole world.

The law against witchcraft in Uganda which was drafted by foreign experts condemns some important aspects of people's traditional heritage.

The response of the population about the threat of extinction of traditional knowledge has ranged from indifference to despair.

Intellectual Property Rights Laws in Uganda

It is important to note the different components of intellectual property rights and how they are covered by Ugandan legal system. The intellectual property rights components include; patent, copyrights, trade marks, designs, geographical indicators, integrated circuits and plant variety.

Uganda has *patent* (Amendment) Act 2002 for **patent**, *copyright and Neighboring Rights Act*, 2006 for **copyright**, the *Trade Mark Act cap 217 laws of Uganda* for **trade**

mark, *(Chapter 218, protection), laws of Uganda 1937* for **industrial designs** and draft *plant variety protection bill* which is presented before parliament.

Uganda has no any law covering **geographical indication** and **integrated circuits**.

Uganda is a member of the following international instruments; ARIPQ, PCT and WTO.

Uganda is not yet a member of Berne, Budapest, Madrid protocol, Strasburg UPOV and Paris Convention.

The government policy support to develop traditional knowledge has included;

- ✰ Restoration of cultural institutions. However a question mark remains because they are disempowered to take on their natural and constitutional roles.
- ✰ Setting up the Uganda national policy on culture which is facing a challenge of implementation due to lack of political will and budgetary support.
- ✰ The policy on use of vernacular language at lower level of instruction (lower primary school level).
- ✰ The development of the draft policy on traditional and complementary medicine.
- ✰ Creation of a policy environment that allows the private investment in traditional knowledge and cultural sector.

At the same time, this support is inadequate;

- ✰ The whole of culture sector has only 0.03 per cent of the Uganda National Budget (Uganda National Budget 2009/10)
- ✰ Culture is only a department in the ministry of Gender, Labor and Social Development.
- ✰ Uganda has no national council on Arts and culture for policy support to the sector.
- ✰ Government has not invested in culture and heritage. There is only one national museum and national theatre built in the colonial era in the Capital City of Kampala.
- ✰ The relevant international conventions like the 2005 UNESCO Convention on protection and promotion of diversity of cultural expression isn't ratified.

This situation reflects government's limited appreciation of the role traditional knowledge and heritage can play in national development. This also shows that the sector itself is not organized and developed enough to package serious demands on the policy agenda and engage government.

The Protection of Traditional Knowledge in Uganda

Uganda's copyrights laws and Intellectual Property Rights stem from the British copyright and ordinance of 1915, the Copyright Act of 1956 and from its intellectual property regime.

Quoting Okello Ogwang (Sillars, Johannessen and Dipio 2009), Uganda's legislation on copyrights and intellectual property rights simply customized the British copyright and intellectual property legal regimes and therefore, unable to cover the non-Western intangible artistic and cultural forms like African Oral Tradition. The legislation accord minimal recognition to local and indigenous cultural forms and place high premium to Western models of authorship and ownership.

The 1964 Copyright Act restricted authorship to only cinemography and written works at the expense of oral tradition and intangible folklore forms. The copyright and neighboring rights 2006, under the section *work eligible for copyrights*, at the end of a long list it remotely mentions "any other work in field of literature, traditional folklore and knowledge."

Unlike in other types of works of Art, the law doesn't present any details on production, presentation and distribution along which protection of traditional knowledge can be operationalised. It is observed for instance that the commercial and especially advertisement sector in Uganda exploit forms of folklore and oral tradition without any legal or ethical restrain.

The traditional knowledge in Uganda is not protected and there are no efforts yet to that effect (James Wasula, 2010). What exist is a one sided debate among the lawyers on whether Traditional Knowledge and folklore can qualify for protection under the copyrights and intellectual property regimes. The lawyers agree that the traditional knowledge doesn't fit in the criteria and standards set to define who can be protected, what can be protected and for how long it can be protected.

The issues that disqualify traditional knowledge for protection under intellectual property principles are as follows;

1. Ownership and authorship. While in copyright the author must be clearly known and identifiable, in traditional knowledge it is more often very difficult to identify the author. The creator is not known (Akubu Jeroline, 2005).
2. While copyright aim at rewarding and encouraging individual creativity, traditional knowledge emphasize the collective interests and ownership.
3. The principle of originality also doesn't stand in the case of traditional knowledge which often is built by cumulative additions by many people of different generations.
4. While in copyright the work must be written, recorded or reduced to material form, traditional knowledge is more often oral and if it is written it can not have been done by original author.
5. Equally, the principle of duration as applied on copyright laws cannot be applied on different forms of traditional knowledge.

The Emerging Questions in the Protection of Traditional Knowledge and Folklore

In attempt to develop protection of traditional knowledge, there are challenging questions that begin to emerge.

What would be the definition of public domain in the context of African traditional knowledge and folklore? Such definition should consider the African traditional systems of authorship, ownership, hereditary and inter-generational transfer of information. Who is the author of a piece of folklore which has existed for generations passed on by word of mouth? A piece of art could be created by imaginations of one person exclusively or by collective contributions. One piece may be an anthology of many contributors.

Who should claim authorship of a dying traditional art piece, the initial author or the one who re-invents, adds value and makes it relevant to today's world? How do we reconcile the need to maintain authenticity and demand to adapt to new circumstances?

What do we do with important forms of traditional heritage being condemned by emerging beliefs and value systems and some forms of traditional knowledge being pirated from one society to another?

Recommendations

It is recommended that first and foremost, efforts should be focused on mobilizing the communities to reclaim the value and vitality of traditional knowledge and re-integrate it in the mainstream development process.

The traditional and cultural institutions in Uganda like the four major kingdoms *i.e.*, Buganda, Bunyoro, Tooro and Busoga, the chiefdoms like Kooki, Rwenzururu, Tien Adhola, Woni Nyachi and traditional clans should be encouraged to focus on their natural mandate of conserving and developing the traditional values and heritage.

The civil society and non-governmental organization as facilitators of development should increase their focus on development of traditional knowledge. The role of civil society of building local and international networks, organizing the sector, generating demand for protection and directing public opinion on developing traditional knowledge is clearly missing.

A programme called Traditional Knowledge Development Programme has been conceptualized by practitioners with a mission of mobilizing communities to engage in a conscious process of organizing, re-inventing, securing, protecting and re-constructing traditional knowledge in the contemporary development. The programme shall coordinate and network practitioners from different components in traditional knowledge like folklore, traditional herbs/medicines, indigenous languages and music. It shall build networks for enhancing collective vision and voice in the sector and to catalyze organization development and protection of traditional knowledge.

A study should be carried out to look into the possibility of developing a unique intellectual property regime for the traditional knowledge in Uganda.

A digital database system of traditional knowledge should be developed in Uganda to safeguard against inappropriate use and piracy of Ugandan Traditional knowledge. Model case studies of such systems in developing countries include;

Traditional Knowledge Digital Library (TKDL), Council of Scientific and Industrial Research, India and National Indigenous Knowledge Systems Department of Science and Technology, South Africa.

References

Akubu Jeroline (2005). Challenges of current copyright regimes in protecting indigenous knowledge and African traditions. Uganda Law Reform Commission. Kampala – Uganda

BBC, 2009. The tragedy of dying languages. Available at: http://news.bbc.co.uk/2/hi/8500108.stm

BBC, 2009. Saving Native American languages. Available at: http://news.bbc.co.uk/2/hi/americas/7964016.stm

Cross Cultural Foundation of Uganda 2008. *Culture in development in Uganda: Experiences and prospects*. Kampala – Uganda.

COMPAS endogenous development magazine 2009.

Interview with James Wasula – Executive Secretary of Uganda Performing Rights Association on 3rd July, 2010 in Kampala - Uganda

Ministry of Labour, Gender and Social development, 2009. Uganda National Cultural Policy. Kampala – Uganda.

Sillars S, Johannessen L, Dipio D (eds.) 2009. Performing Change. Novus Press, Oslo.

Uganda National Budget 2009/2010

Uganda Copyright and Neighboring Act, 2006.

Chapter 9
Intellectual Property

Frederick Okumu
Network Officer,
Uganda Industrial Research Institute (UIRI),
Kampala, Uganda
E-mail: fredokum@hotmail.com, okumu@uiri.org

ABSTRACT

There are many existing programs and initiatives in Uganda supporting small businesses, but tend to suffer from a number of weaknesses. In particular small businesses find it difficult to do research and development, commercialize their results in markets (innovation) as fast as they should. A business incubator is an organization that supports the creation and growth of new businesses by providing services and infrastructure that is required by the targeted clients. Intellectual property rights management is geared towards the development of research findings into innovations, transfer thereof to users and private sector aimed at their commercialization and exploitation of the product findings.

For micro, small and medium enterprises to be dully competitive in a competitive economic environment requires that they develop internal capabilities to effectively assimilate, use and adapt product and process technologies for their businesses to survive on an ongoing basis.

To overcome this drawback, the concept of Business Incubation has been proposed. The key idea is to create and nurture new businesses for growth by providing services and infrastructure required by utilizing the external knowledge sources (open innovation) model which assists formation of business and industrial clusters.

Given that most firms in developing countries (Uganda) start too small to compete especially in international markets, a pre-requisite to industrial development, governments and policy makers should give particular attention to the constraints and needs of Micro, Small and Medium Enterprises. This can be done by adopting a mix of policies and framework conditions to reduce on the obstacles that hamper technological innovation, collaboration and business growth. In particular Intellectual property rights management should be managed by institutions that are near to people (like its done in India) other than these so called big names (Government ministries).

Introduction

Intellectual property plays a significant role in all aspects of science and engineering as evidenced by the intellectual property market, which continues to grow at a remarkable rate.

With regard to patents, the rules for applying and qualifying for patent protection are changing as quickly as the technology they protect advances.

Now, more than ever, scientists, engineers, managers and the like must understand their companies' intellectual property assets and their alternatives for realizing a significant return on investment for these valuable assets.

Moreover, global markets require an understanding of how intellectual property can be protected and exploited worldwide.

Types of intellectual property - patents, trademarks, copyrights, trade secrets and their importance to the success of technology-based companies.

The Case for Uganda

Ugandans practicing scientific and engineering professions do not have any knowledge on whether to patent their products or not. They believe patenting is only for the western world that has the idea of innovation and as made everything. Scientifically the technology aspect is not being protected at all.

It's the idea of Uganda Industrial Research Institute (UIRI) to bring in the IPR system into action, now that UIRI is to spear head the industrial and technological development of Uganda.

Why this Concept

An invention must meet several criteria if it is to be eligible for patent protection. These include, most significantly, that the invention must consist of patentable subject matter, the invention must be industrially applicable (useful), it must be non-obvious, and the disclosure of the invention in the patent application must meet IPR standards, This makes IPR difficult to exist in the Ugandan culture.

For Uganda the patent laws are not yet in full use. For the Ugandan case, it's only the musicians who are trying to force the law into action to protect their music creations.

Understanding the scope of an invention described in a patent, this many (SMEs) feel is time wasting and therefore they rather do without the patent.

The Problem

Lack of adequate human resources to publicise IPR. The use of the IP system by SMEs in African (Uganda) region is not well documented. However, general indications seem to suggest that African SMEs have a low level of awareness of the potential of the IP system, and understandably use it very little or not at all. Basically one can identify two main reasons for low use of the IP system by SMEs in the African region, SMEs have to look at IP not as a mechanism to prevent others from taking advantage of a technology or entering a market but rather as a very useful business tool.

Most SMEs use trade secrets as business strategy some of them not knowing that such secrets can be legally protected.

One of the guiding principles of Uganda Industrial Research Institute (UIRI) operations is the deliberate attempt to engage in sustainable projects, or at least to assure sustainability of any projects to engage in. Prof Charles Kwesiga always says "... having joined the business incubation bandwagon we quickly developed the modalities of our incubator practice and modified the terms of reference for dealing with partners so that they could be part of our incubator portfolio".... UIRI embarked on establishing two types of incubators: in-house incubation and virtual incubation, with the latter taking centre stage because UIRI already had a partnership with budding entrepreneurs in peri-urban areas, away from UIRI campus.

UIRI's business incubation portfolio has grown to include such activities and products:

- Software for biometric solutions
- Design and assembly of equipment for teaching science (e.g. signal generators, power supply units)
- Manufacture of vaccine against Newcastle disease in poultry.
- Food processing technologies, adding value before the table

Many more projects are in the pipeline and the Government of the Republic of Uganda has come out in full support of UIRI's incubator initiative. Uganda Industrial Research Institute if given the mandate of IPR awareness in Uganda, it can reach many business organs that would benefit from the IPR system given the level of their business penetration strategy.

Objectives of the Incubation Program

- Reduce the failure rate of start-up enterprises
- Develop profitable enterprises that can create higher incomes to the entrepreneurs and the opportunities for offering employment
- Support government efforts in implementing strategies for poverty reduction
- Enhance collaboration between research institutions, academia, and private sector (SMEs)
- Assist the government in refining policies, particularly in relation to SME development
- Increasing the IPR awareness in doing business
- Establish the requisite networks between enterprises and other supporting agencies

HE President Museveni together with Dr. Nantulya of Vaccine Project

HE President Museveni together with Dr. Mwanje of E-Tech in the Electronics Workshop

High Quality Meat Products from Gourmet King

A-Tech Software Developer (Biometrics technology)

Figure 9.1: Highlights of UIRI's Business Incubation Programme

Conclusion

Developing countries should stop coming up with bills which will only be shelved and will help the poor man or woman who is trying to protect their idea. Africa in particular is politicising, only the ruling group of people benefit from all sorts of ideas that come up. We in Africa still a lot of innovation ideas that can change the world and make it a better place.

Let's encourage Intellectual property rights publication in all sorts of areas; this will encourage young innovators to open up. There should be a practical consideration of intellectual property that relates to the management of intellectual property, which involves the development of research findings into innovations and inventions and subsequently transfer thereof to users and the private sector with a view to commercialization and exploitation.

For God and my Country.

Literature Consulted

Literatures and lectures by Prof. Charles Kwesiga, Executive Director, Uganda Industrial Research Institute.

The Role of Intellectual Property System in the use of ICTs by SMEs by Christopher M. Kalanje (Consultant-WIPO).

Chapter 10

The Scientists and IPR in a Resource Starved Nation: NIPRD IPTTO Experience

Emeje, Martins and Gamaniel Karnyius
National Institute for Pharmaceutical Research and Development (NIPRD), Idu, Abuja, Nigeria
E-mail: martinsemeje@yahoo.com; chubiindia@gmail.com

ABSTRACT

In 2006 the Federal Ministry of Science and Technology (FMST) through the National Office for Technology Acquisition and Promotion (NOTAP) established Intellectual Property Technology Transfer Offices (IPTTOs) in 15 selected R&D institutions (10 universities, 4 research institutes and one polytechnic) in the country. To provide training for the newly appointed Heads of the IPTTOs, NOTAP in collaboration with the World Intellectual Property Organization (WIPO), organized two workshops in Abuja in which the IPTTO heads were exposed to the principles and management of the intellectual property rights, licensing of intellectual property, innovation promotion and development of research-industry linkages. After the workshops a study tour for the newly appointed Heads of IPTTOs to the various models, structures, and operations of TT Offices in universities and research institutions in the UK was undertaken. NIPRD was part of it. Three years later (2010), not much have changed, the reasons for this is presented in this chapter.

Keywords: *Nigeria, Intellectual Property NIPRD, IPTTO.*

Introduction

Intellectual Property Rights in Nigeria

Intellectual property covers copyrights, trademarks and patent laws (Richard and Roberto 2006). Trademark law protects the identity of a brand, copyrights preserves the rights to creative works, while Patent law safeguards the rights over scientific/technological innovations. In event of infringement, a patent owner can get legal redress either by stopping the practice or by suing for damages. The Patent Act of 1990 is the governing Patent law in Nigeria and prescribes if and whose product may be granted the statutory rights. S.344 of this act lays down rules that determine if a product Patented in favor of the inventor.

This law states that an invention is Patentable if:

- It is new: Section 1 (2) of the act states that it will qualify as new when it does not form part of the state of the art. The term "state of the art" means to that field of knowledge that has been made available to the public anywhere and anytime whether by oral, written description or by public use. However this invention shall not be regarded as having been made available merely because a person exhibits to the public his own invention within a period of six months preceding the filing of a patent by another inventor. This rule is adhered to whether or not it has been exhibited in an officially recognized international exhibition.
- Resulted from inventive activity and is capable of being used in an industry.
- Constitutes an improvement on a patented invention.

The rights to a patent are vested in the "Statutory Inventor" who is simply the first person to file and register their patent and is legally credited with being the inventor. However, Section 2(2) of the act provides opportunity for redress if there was "fraud" by reassigning the rights to an invention, to a person who is adjudged the true inventor whether or not he is the first to register a product.

Getting a Patent

Applications for a patent are usually made to the Registrar of Patents and Copyrights. A search is then conducted to ensure that the product has not already been patented. After the search has been completed, and the product is confirmed not to have been registered yet, the inventor is free to go ahead and proceed with registration.

An application can then be made to the registrar and must contain;

- The applicant's full name and address.
- Description of the relevant invention along with any plans and drawings.
- A claim that defines the particular protection or rights sought after by the applicant.
- Such other matter as may be prescribed depending on the facts of the application.

These requirements will be accompanied by:

1. A prescribed fee.
2. Where appropriate, a declaration by the true inventor requesting that he be mentioned as such in the Patent with his name and address.

3. In event of where a legal practitioner makes the application on behalf of the applicant, a signed Power of Attorney will be required.

Once awarded the Patent rights the inventor is free to exercise his rights over it as he may choose. He may if he chooses, decide to assign his proprietary rights over the invention to any person he wishes in any geographical territory. He may also upon consideration or payment, grant a Patent license to another party on terms agreeable to him.

Establishment of IPTTOs in R&D Institutions

In 2006 the Federal Ministry of Science and Technology (FMST) through the National Office for Technology Acquisition and Promotion (NOTAP) established Intellectual Property Technology Transfer Offices (IPTTOs) in 15 selected R&D institutions (10 universities, 4 research institutes and one polytechnic) in the country. The establishment was considered imperative because of the following facts:

1. Research work already carried out by researchers were not being patented, nor exploited, licensed or commercialized.
2. Most research activities are not coordinated institutionally.
3. Most research activities are not linked to industry, nor are entrepreneurs attracted to utilize the potentials therein.
4. Many are not targeted to national priority, national needs, nor used to address pressing needs.
5. Many useful results are not being brought into national prominence, nor made visible to potentially interested parties.
6. Substantial components of the intellectual capital of the institutions are not being invested and are virtually idling away.

The envisioned mandate of the IPTTO in each institution included the following:

1. Patenting all innovations
2. Refocusing R&D projects
3. Linking R&D to industry needs, i.e. build research-industry linkage
4. Coordinating research
5. Promoting facilities and funding for R&D
6. Marketing R&D results
8. Creating technology-based SMEs from High-tech start-ups, and spin-offs or spin-outs
9. Boosting innovation and creativity among the populace especially, the youth.
10. Generating entrepreneurialism among researchers and students, and thus help diversify the economy.

To provide training for the newly appointed Heads of the IPTTOs, NOTAP in collaboration with the World Intellectual Property Organization (WIPO), had organized two (2) workshops in Abuja (August and November, 2006) in which the IPTTO heads were exposed to the principles and management of the intellectual property rights, licensing of intellectual property, innovation promotion and development of research-industry linkages. After these two workshops it was necessary to provide a hands-on experience to the coordinators so as to give them capability to operate and manage innovative, vibrant and successful TTOs that will churn out, innovative technologies to transform the Nigerian economy. To achieve this objective, a study tour for the newly appointed Heads of IPTTOs to the various models, structures, and operations of TT Offices in universities and research institutions in the UK was undertaken. NIPRD was part of it. Three years later (2010), not much have changed.

The brief description of intellectual property in Nigeria above seems good enough for adequate protection of the rights of an inventor. But, does an inventor in Nigeria realize his invention?, what if he does, why does he need a patent instead of publication?, there are many questions. We have recognized the following as the major obstacles to effective IPR in our Institutions.

Lack of Political and Financial Support from Government

Many especially from developed countries argue that government has no business establishing and supporting research programs. In fact, many political leaders in developing countries today have mortgaged R&D in the name of foreign investment. The argument of those in favour of private sector driven R&D stems from the fact that, government support runs counter to international treaties regarding the rules of the game under the WTO. Secondly, that government makes bad judgments when they try to help particular sectors. However, while the ground rules under the WTO inhibit the subsidy of specific commercial products or firms, they do not constrain broad support of R&D and training tailored to meet a sector's needs. It is important for us to know that government support for IPR in some developing countries like India and Brazil have helped to strengthen it. It is therefore advocated that, developing countries under an umbrella body like the Centre for science and technology of the non-aligned and other developing countries programs, makes it mandatory for some substantial level of government support. Some examples of government's support contributing effectively to economic progress include the agricultural and industrial development in Japan in the early stages of Japan's successful catch-up experience, Korea and Taiwan in the last decades of the twentieth century, and at the successful programs in Brazil to support agricultural development and aircraft design and production. We have deliberately ignored mentioning the early stage support and current huge government grant allocated to university research in the United States and Europe which has been contributing to technological advancement in those countries. We are of the opinion that our government needs to do more than setting up agencies and/or committees.

Lack of Support for IPTTOs in the Universities and R&D Institutions

Although the Nigerian government through its responsible agency, NOTAP committed colossal amount of resources to setting up IPTTOs and training (national and international) coordinators, most of these offices are virtually not doing anything. Some of the managers of the institutions (Vice Chancellors and Directors General) consciously or otherwise do not give any attention to these offices. The world intellectual property organization's university IP initiative provides that, the following shall be the duty of the national IPO towards the IPTTOs; Designation of a focal point within IPO, provision of information, for example on national legislation texts and updates, awareness material, methodological support, list of local patents, list of local IP related NGO's and continuous national training. In our case, only the first has been achieved, so how do you expect IP to make useful progress in our universities and R&D institutions? With this picture, can we say that there is any support for these offices? Certainly not!

Lack of National Master Plan for IPR Activities

There is generally, no national master plan for IPR. A national master plan should consist of National IP legislation, strategy and policy, R&D strategy and economic framework. Hear what the Pakistan Press International Information Services said in 2007 when the Pakistan government embarked upon a deliberate attempt to develop a national IPR master plan: "Every nation that broke off the trail of poverty first developed strong intellectual property creation and utilization" the lack of national IP policy we believe is one of the reasons why institutions are not under any obligation to develop their own IPR policies which will incorporate Incentive and Reward system and explicitly state the sharing of royalties and other fees accruing from Technology Transfer activities. Our Institute (NIPRD) for example has only in 2010, set up a committee to develop its local IP policy. We also opine very strongly that inadequate attention to IPR issues in our academic curriculum is one of the militating factors against IP in Nigeria. Many scientists get to hear or know about IPR after graduation. We believe that if IPR becomes part of our academic curriculum, then better progress could be made.

Lack of Functional Relationship between R&D Institutions and Industries

In Nigeria this is a major problem, and we believe that, this type of meeting can assist to curb the menace of "over importation" of cheap products across countries. For example, why is it cheaper for a Nigerian pharmaceutical industry to import drugs from India and China than to manufacture the same drug in Nigeria? The negative effect of this on IPR is obvious; the Nigerian industry will not waste its resources on R&D in Nigeria when it can easily import same products from other countries and at cheaper rates. The following article published in Europolitics 2007, summarizes the importance even developed countries attach to R&D and Industries relationship:

"A set of recommendations has been published by the European Commission, on 4 April, in order to improve the transfer of knowledge between universities and industries. A major point made in the recommendations is that national authorities need to make it easier for staff exchanges to occur between universities and the private sector. *Rules are criticized which, in some countries, prevent public sector researchers from working for industry on a part-time or consultancy basis. Such mobility is important, the Commission believes, in order to enable research institutions to better understand the needs of industry.* Member states are also urged to actively promote the pooling of intellectual property between universities. This is especially important for those which are too small to have their own knowledge transfer offices. These offices are used to establish links with industry to exploit intellectual property rights (IPR). It is noted that the rules governing the ownership of publicly funded research results vary across Europe and that the Commission may decide to examine, in the near future, the question of a single European ownership model. An area identified as requiring special attention at member state level is the promotion of interactions between universities and SMEs. At the moment, most collaboration tends to involve large firms. Good practices highlighted in the paper include innovation vouchers' given to SMEs in the Netherlands (worth €7,500 each), which allow them to buy knowledge and consultancy services from universities. It is recommended that other countries should adopt similar initiatives, including the use of R&D tax credits. The EU executive outlines measures it will take. These include the setting up of an expert group, in 2007, to draw up an EU-wide scheme to measure and compare the universities' innovation-related activities. It is also expected that the planned European Institute of Technology will make a major contribution to promoting cooperation. The communication aims to promote discussion on how to address deficiencies in the way Europe exploits its public R&D. It notes that while the importance of knowledge transfer to the EU's competitiveness is increasingly recognized, many barriers remain. These include cultural differences between the business and science communities, a lack of incentives to cooperate, legal barriers and fragmented markets for knowledge and technology. To help remedy this, the Commission has also published a set of voluntary guidelines for universities and other research institutions to follow to improve their links with industry. These contain recommendations to help universities adopt industry-friendly IPR policies, and guidance on how to handle other contractual arrangements. The Commission sees this as only a first draft of the guidelines, and intends to set up a high level group in 2007 to help improve it". This article was culled from Europolitics 2007 with the expectation that developing countries will learn from the bold decisions taken above.

Conclusions

Assisting research institutions in Setting IP infrastructure (IPTTOs) alone is not enough. There should be a deliberate education of policy makers and R&D managers on IP issues including the role of government in IP Assets management capacity building, National IP Policy, Institutional IP Policy and procedures etc. We also propose that developing countries should collaborate in all areas of IPR with a view to developing perhaps a single or uniform model that will give them ownership of their inventions.

References

Education- HEC for secured int'l patents in research Pakistan Press International Information Services 452 words 13 April 2007 English (c) 2007 Asia Pulse Pty Limited.

Richard R. Nelson and Roberto Mazzoleni (2006). The Roles of Research at Universities and Public Labs in Economic Catch-up. Laboratory of Economics and Management Sant'Anna School of Advanced Studies. http://www.lem.sssup.it/

RESEARCH AND DEVELOPMENT: DISCUSSION LAUNCHED ON PROMOTING KNOWLEDGE TRANSFER 476 words 11 April 2007 Europolitics 3284 English Copyright 2007 Europe Information Service All Rights Reserved www.europolitics.info.

Report of technology transfer study tour to the United Kingdom by the heads of Intellectual property and technology transfer offices in Universities and Research Institutions, and officials of the federal ministry of science and technology: 5 -9th February, 2007; organized by the National Office for Technology Acquisition and Promotion (NOTAP), Abuja, Nigeria.

WIPO Regional Workshop on Intellectual Property and Technology Management for Universities Abuja, Nigeria May 20 – 22, 2008.

Section II

American Continent

Chapter 11

The Intellectual Property Rights in Mexico: The Tequila Case

Ana Paula Ramírez Ramírez

Tequila Regulatory Council,

Av. Patria 723, Jardines de Guadalupe, Zapopan, Jalisco, México

E-mail: ana.ramirez@crt.org.mx

ABSTRACT

The present chapter is divided in two sections: in the first one one will find a compilation of the International and Bilateral Agreements Mexico has signed related with Intellectual Property Rights Protection, also Then one will briefly learn about the policy of Mexico regarding Intellectual and Industrial Property Rights protection, by analyzing the structure and the way this protection is managed by the different agencies of the government in a coordinated manner, involving areas of specialization which are quite diverse all inter linked in the aim of getting the Intellectual Property Rights protection assured from different perspectives. In the second section, the focus is made on the study of Geographical Indications, and the way they are protected in general in the World and in Mexico in specific, to pass then on to the subject of the Appellations of Origin of Mexico and their general situation, to continue finally with the Tequila case exploring its particulars, its regulatory council, the infringements against Tequila and the way they are faced, and we will end with the actual Law Initiatives there are in place now and that would very much improve the Intellectual Property Rights protection of the Appellations of Origin in Mexico.

***Keywords**: Intellectual Property Rights, Mexico, Appellations of Origin, Tequila*

Introduction

Mexico is a New World Country but when it comes to the Intellectual Property Rights (IPR) Protection it has learned fast. The issue has been taken seriously, and in order to prove that, the country has many different international Intellectual Property Rights IPR Agreements signed, Mexico is also part of some of the most known movements for enhancing the fast protection of IPRs worldwide today.

The idea of protecting traditional goods such as Appellations of Origin (AO) was brought from Europe as the local history tells. The recognition of Appellations of Origin in Mexico startedin 1973, when the modality of AO was added to the Industrial Property Law; the next year in 1974, Tequila was recognized as the First Appellation of Origin, and from then on 12 additional Appellations of Origin have been protected. The Mexican State in its different governmental levels do believe the AOs will bring development and well being to the inhabitants of the region that may produce them,. This is one of the reasons why producersget organized to ask for the recognition by the Mexican State; however, in order to get those benefits, it is important to support the value that the specific good already has and to implement control mechanisms for that, that is to have proper standards for each product and mechanisms to ensure its compliance; and then, once the AO is strong and stable it will definitely need to be protected from counterfeiting and in issues of infringements, authorities are to be enabled to execute exemplary actions in order to really dissuade the offenders from repeating the infringement.

The issues to be tackled with regards to the AOs best use are:

- ✰ Adherence to minimum compliance requirements (to ensure a continous quality)
- ✰ To count on a strong legal framework that defends the misuse of a national Intellectual Property asset such as Appellations of Origin.

The legal initiatives to get the latter in place are already embarked upon and currently in Congress and in the Senate of Mexico, it is just a matter of will by the legislators to let them succeed and by implemented and then the Mexican IPR system for protecting Appellations of Origin will be much stronger and with that the Appellations of Origin will grow stronger too bringing economic development to the producers of AOs.

International Legal Framework on Intellectual Property of Mexico

International Treaties and Agreements Related with Intellectual Property Rights Protection Signed and Ratified by Mexico

Mexico is part of different international agreements related with the protection of Intellectual Property Rights (IPR´s). In this section a list of the agreements signed by Mexico is given, as well as the year when they were in force. As a member of the World Trade Organization, Mexico is part of the Agreement of the Trade-Related Aspects of Intellectual Property Rights TRIPS (in force from January 1, 2000) which is the WTOs most important IPR Treaty that protects GIs.

Mexico is part of the following IPR Treaties which are managed by the World Intellectual Property Organization (WIPO), sorted by date of issuance:

- ☆ Paris Convention on the Protection of Industrial Property of 1883; (Mexico is part of the Stockholm Act of 1967, since July 26, 1976).
- ☆ Berne Convention for the Protection of Literary and Artistic Works of 1886 amended last in 1979 (in force from June 11, 1967).
- ☆ Nice Agreement Concerning the International Classification of Goods and Services for the Purposes of the Registration of Marks of 1957 (In force from March 21, 2001).
- ☆ The Lisbon Agreement for Protection of Appellations of Origin and their International Registration of 1958 (Mexico is part of the Lisbon Act of 1958 since September 25, 1966; and of the Stockholm Act since January 26, 2001).
- ☆ Rome Convention for the Protection of Performers, Producers of Phonograms and Broadcasting Organizations of 1961 (in force from May 18, 1964).
- ☆ Convention Establishing the World Intellectual Property Organization of 1967 (in force from 1975).
- ☆ Locarno Agreement Establishing an International Classification for Industrial Designs of 1968 (In force from January 26, 2001).
- ☆ Patent Cooperation Treaty (PCT) and its Regalement of 1970, last modified 2001 (in Force from January,1, 1995).
- ☆ Strasbourg Agreement Concerning the International Patent Classification of 1971 (In force from October, 26 2001) and is also part of the IPC Union Assembly.
- ☆ Convention for the Protection of Producers of Phonograms Against Unauthorized Duplication of Their Phonograms of 1971 (in force from December 21, 1973).
- ☆ Vienna Agreement Establishing an International Classification of the Figurative Elements of Marks of 1973, amended 1985 (In force from January 26, 2001).
- ☆ Brussels Convention Relating to the Distribution of Programme-Carrying Signals Transmitted by Satellite of 1974 (In force from August 25, 1979).
- ☆ Budapest Treaty on the International Recognition of the Deposit of Microorganisms for the Purposes of Patent Procedure of 1977 (In force from March 21, 2001).
- ☆ Nairobi Treaty on the Protection of the Olympic Symbol of 1981 (in force from May 16, 1985).
- ☆ WIPO Performances and Phonograms Treaty (WPPT) of 1996 (in force from May 20, 2002).
- ☆ WIPO Copyright Treaty of 1996 (In force from March 6, 2002).
- ☆ Finally, México is part of The International Union for the Protection of New Varieties of Plants (UPOV) from 1961 (in force from August, 1997).

What can be inferred from the number of treaties ratified by the Mexican Senate is that the Government of Mexico is interested in supporting the Intellectual Property

Rights Protection which is explicitly stated in the National Development Plan and is focused in solving some challenges through infrastructure improvements and legal modifications that could be made to improve the national system, as we will learn in the next sections.

Free Trade Agreements that include an Intellectual Property Chapter, Signed with Different Countries by Mexico

Mexico is the country with the most Free Trade Agreements signed with the World. Mexico has undersigned 12 Free Trade Agreements with different countries and regions, which adds up to 44 countries with which Mexico has better trade conditions; this fact has turned Mexico in the only Latin-American Country having free trade agreements with the main economies of the world.

In all of these agreements a chapter or some disposition about intellectual property is included, the mutual protection for the national GI´s or distinctive products are part of those IP dispositions:

- ✰ Agreement between the European Community and the United Mexican States on the mutual recognition and protection of designations for spirit drinks, 1997. Mexico recognizes 229 spirits of the UE and the UE recognizes 4 spirits from Mexico with Appellation of Origin.
- ✰ North American Free Trade Agreement/NAFTA): Canadá, The United States and Mexico (Part VI, Chapter XVII is on Intellectual Property) Signed December 17, 1992, in force from January 1994.
- ✰ Free Trade Agreement between Mexico and the Republic of Costa Rica (Chapter XIV is on IP), in force from April 1995.
- ✰ Free Trade Agreement with Colombia (Chapter XVIII on IP) In force from 1995.
- ✰ Free Trade Agreement with the Republic of Bolivia (Chapter XVI on IP), in force from 1995.
- ✰ Free Trade Agreement with the Republic of Nicaragua (Chapter XVII on IP), in force from 1998.
- ✰ Free Trade Agreement with Chile (Chapter XV on IP), in force from 1999.
- ✰ Free Trade Agreement with the European Union (Chapter V on IP) in force from 2000.
- ✰ Free Trade Agreement with the State of Israel, in force from 2000.
- ✰ Free Trade Agreement with the European Free Trade Agreement EFTA (Switzerland, Norway, Island and Liechtenstein) (Chapter VI on IP) in force from 2000.
- ✰ Free Trade Agreement with the Republics of El Salvador, Guatemala and Honduras (North Triangle), (Chapter XVI on IP). In force from 2001.
- ✰ Free Trade Agreement with Uruguay, (Chapter XV on IP) in force from 2004.

- ☆ Agreement for the Strengthening of the Economic Association with Japan, in force from 2005, in this Agreement the cooperation regarding IP issues is considered. An additional Agreement in the framework of the 2005 Agreement was signed in 1998 for protecting national GIs mutually, Mexico recognizes 3 Japanese GIs: Iki, Kuma y Ryuku; and Japan recognizes the Mexican Appellations of Origin: Tequila and Mezcal.

Policies of Intellectual Property Rights Protection of the Mexican Government

There are different policies implemented by the Mexican Government which include the IPR Protection as an important fact for reaching an efficient development of the Nation in general. The Presidency recognizes that it is crucial to count with certainty regarding the property rights in order to benefit the whole economic activity, since this allows the economic actors to have real incentives for innovating and investing in development of solutions that are not yet available.

In the National Development Plan of the Actual Presidency of Mexico (PND) 2007-2012, it is written that the rule of law allows the social processes in all spheres of life to unfold with efficacy.

In this Plan it is considered fundamental to establish the standards that will allow legal certainty and render the application of the Law predictable. Regarding the intellectual property rights in such Plan it is stated that an extended infringement is the violation of the intellectual property through piracy and that this would discourage innovation and destroy the actual and the potential entrepreneurs motivation, adding that the national laws and international agreements regarding the protection of copyrights and patents will be enforced, by means of merchandise seizures and preventive actions at Customs, to stop the entrance of illegal products into the Country.

Furthermore, the "Ministry of Economy Program by Sectors 2007-2012" establishes in its guiding objective "2.6. To provide legal certainty to the economic factors through the enforcement and adjustment of the legal framework, as well as through the modernization of the processes for opening enterprises", and in order to reach that goal it notes different strategic lines such as "2.6.5 To provide protection to the industrial property rights", "2.6.6 To fight smuggling and piracy" and "2.6.7 To prevent and fight the actions of untrustworthy competence related with intellectual property". In the Program it is said that part of the actions that will be carried out for the fulfilment of such strategic lines, is to provide legal certainty to the legitimate owners of an invention or a distinctive sign through an adequate attention to the protection requests the Ministry receives.

The culture of industrial property rights protection and more specifically, the patents system has not been very much used by the Mexican science and technology research centres, resulting in a disadvantage position for them, since it is an indispensible tool for the promotion and fostering of science and technology of any country. Thus, the National Council of Science and Technology Council (CONACYT) has been working together with the IMPI in order to include the IP subject as part of the operation rules of the different aid programs of CONACYT. In June 2009 a Reform to the Science and Technology Law was published in the Official Gazette (DOF, for its Spanish Acronym) through this Law CONACYT creates a body to coordinate the research policies and amongst its powers is to approve and create policies and mechanisms to support science and technology, amongst them the ones related with the IPR framework.

Structure of the IP System and Strategies of the Government to Promote and Create Awareness of IPR's, to Protect and Fight IPRs Infringements

The Institutional framework of the industrial property system of Mexico is formed by various entities of the centralized and decentralized public sector, being the main entity the Mexican Institute of Industrial Property (IMPI, for its Spanish acronym).

The IMPI is a decentralized organism, with its own patrimony and legal personality, it was created in 1993 through a Presidential Decree and is in charge of managing the industrial property system of Mexico and is one of the management authorities in charge of monitoring the compliance with the industrial property rights, together with: the Specialized Office of Prosecution of Intellectual Property of the Attorney General of the Mexican Republic (PGR for its Spanish Acronym); and the Ministry of Agriculture, Cattle, Rural development, Fishery and Food; and the Mexican Treasury. Regarding civil and criminal matters the competent authorities are the District Juries and the Circuit Collegiate Courts.

The IPR Protection System in Mexico involves many different agencies and the guidance is in charge of the IMPI which is a decentralized public organism dependent of the Executive Power of Mexico and grouped in the sector leaded by the Ministry of Economy. IPR protection for Mexico is a vital tool for reaching development, as is stated by the IMPIs, Director General, Jorge Amigo Castaneda, in its welcoming words to the IMPIs Webpage:

"The IMPl objective is to protect the industrial property rights and promote and spread the benefits it offers in order to support the inventive and commercial activity of our country, fostering the creation and development of new technologies in the benefit of all society."

Through these words the general overview of what IPR protection means to the country is established. It is a tool for development and the better it is managed the more benefits society will get.

The authorities that manage the IP system in Mexico are summarised in the Figure 11.1.

The Mexican Institute of Industrial Property (IMPI).

The administrative bodies of the IMPI are the Government Board and the Director General. The Government Board is headed by the Minister of Economy and includes the following Ministries:

- ✰ Ministry of Economy
- ✰ Ministry of Treasury (Tax and Revenue)

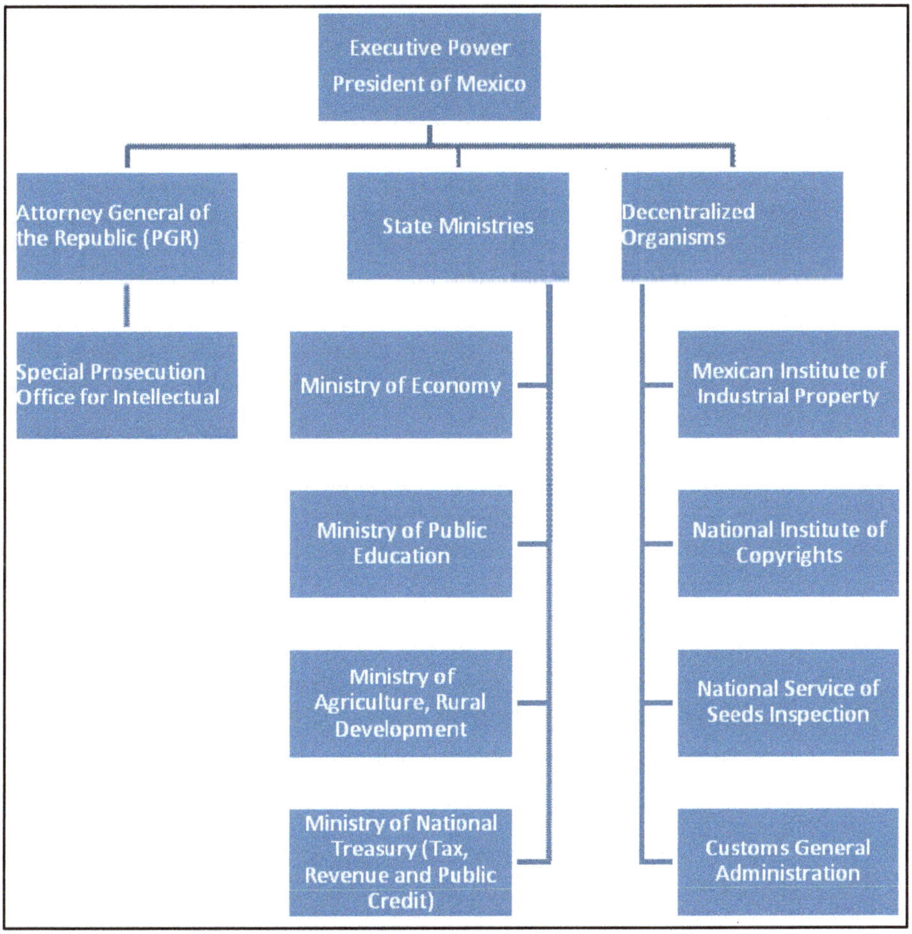

Figure 11.1: Authorities that Manage the IP System in Mexico

- Ministry of Foreign Affairs
- Ministry of Agriculture, Rural Development, Fishing and Food
- Ministry of Education
- Ministry of Health
- National Council of Science and Technology
- National Centre of Metrology

The objectives of competence of the IMPI are:

- To foster and protect the industrial property rights;
- To prevent the acts that infringe against industrial property or that are acts of disloyal competence, and to establish the corresponding penalties and sanctions.

- To promote and foster the inventive activity of industrial use, the best techniques and the spread of technological knowledge in the productive sectors.

Therefore, the main functions of IMPI are the following:

- To process and in its case, grant invention patents and utility model registers, industrial designs, trademarks and commercial ads, to issue declarations of protection to appellations of origin, to authorize the use of them; to publish the commercial names, and to register their renovations, transmissions or licences of use and exploitation.
- To substantiate the nullity, expiration and cancellation of the industrial property rights procedures, to elaborate the resolutions and to issue the due administrative declarations.
- To carry out the investigations about alleged administrative infringements, to order and practice inspection visits; to require information; to order and execute the provisional measures in order to prevent or make cease the violation to the industrial property rights; to hear the alleged offenders, and to impose the due administrative sanctions regarding industrial property.
- To promote the international cooperation through the exchange of administrative and legal experiences with institutions in charge of the register and legal protection of industrial property in other countries.
- To carry out studies about the industrial property situation in the international arena and to take part in the meetings and international for a related with the matter.
- To carry out the legal publication, through its Gazette, as well as to spread information about granted patents, registers authorizations and publications.
- To sanction the infringements about commerce regarding Copyrights.

The competence of IMPI is federal, it covers the whole national territory and it action of competence is in the administrative level, however the IMPI may act as referee in controversies resolution related to the payment of damages (through the civil channel) derived from the violation of the industrial property rights which are ruled by the Industrial Property Law, and may serve as referee when the involved parties name IMPI expressly as such.

Strategies to protect Intellectual Property Rights

The IMPI has been leading in Mexico the awareness actions about the importance of IP rights, promoting the benefits and the possibilities a right holder has to prevent infringements against his intellectual or industrial assets, amongst these strategies to protect IPRs can be noted different activities which include, but are not limited to the following:

Registration Services

Given that one of the IMPIs objectives is to set the basis for the protection of the industrial property rights it is part of its responsibilities to process the applications from the public to register their IP assets: trademarks, collective marks, patents, industrial designs, commercial ads, utility models, and appellations of origin. The IMPI has been developing its personal and technological infrastructure throughout the last 15 years in order to better serve the public and respond faster to the users of the registration services.

From the IMPI services that are offered to the public, the one that is most used in Mexico today is the Trademark Registration, in the years 2007 and 2008, the IMPI received 78,581 trademark registration applications and 71,561 in the year 2008.

In the other hand, in order to get a better response to the patents applied for the IMPI grew its staff for this area which allowed for more resolutions of patents requested, reaching 10,493 favourable in 2008; other service were the utility models which reached 216 granted and the industrial designs that reached 3,144.

Promotion and Education about the Industrial Property System-Awareness

The IMPI has 5 regional offices apart from the Mexico DC one, that cover the whole Mexican Territory, through these offices the IMPI carries out a series of conferences, workshops, courses, expos and fairs, all focused in promoting the IP system and its benefits. From the year 2006 to 2009 the total of promotion activities added up to 2,830.

Many of these activities are carried out in conjunction with the Academy, be it the local Universities or the WIPO Academy for promoting and supporting with part of its staff in the online courses the WIPO offers. It is important to say that from 2007 to 2009 the IMPI has established cooperation agreements with 15 of the most prestigious Mexican Universities and Technological Institutes which has lead to establish a Patenting centre in the universities where the public may get advice regarding the inventions registration and protection.

Compliance of the IP Law

Piracy and IPR infringements in general are fought in a coordinated manner amongst different authorities of Mexico, the leading agency for this activity regarding infringements if the Attorney General of the Republic (PGR), the agency for health matters is the federal Commission for the Protection against Sanitary Risks (COFEPRIS), the Federal Police is the one in charge for the flagrant infringements and for supporting the other authorities, and the Ministry of the Treasury, Tax and Revenue together with its dependent agency of Customs are that charge the fines and keep the illegal merchandise at the borders.

The IMPI, in the other hand is the national authority in charge of the registrations, patents and trademarks granting, and able to declare nullity or expiration of the registers when a third party is affected by a granted registration, also it may impose sanctions for the wrongful use of intellectual property rights. The IMPI has the responsibility to verify the compliance of the IP Law and of the Copyright Federal Law, therefore it does inspection visits, these may be as per a third party request or just through officio visits. From 2000 to 2009 the IMPI reports to have done 11,704 third party request inspections, and 22,443 officio inspections. From these inspections the IMPI was able to seize from 2000 to 2009, 40,586,803 products with a value estimated in USD 15,640,630.79.

Given the importance of the piracy problem the IMPI has created a new division made of specialists focused in doing investigation in order to detect counterfeit merchandise and issue reports with the statistics on the results reached and who are in charge of coordinating the different authorities at national and international level, involved in the IP protection through cooperation agreements, education events, and research projects, inspections visits, amongst other coordinated activities such as public awareness and campaigns against piracy.

Situation of the Intellectual Property Rights of Geographical Indications in Mexico and the Appellation of Origin of Tequila Case

What is a Geographical Indication?

Its intellectual property rights and ways to defend them internationally.

Geographical Indications

"A Geographical Indication (GI) is a sign used on goods that have a specific geographical origin and possess qualities, reputation or characteristics that are essentially attributable to that origin. An appellation of origin (AO) is a special kind of GI" and we are going to be focusing in this since Tequila is an AO.

It is important to emphasize that the product with GI or AO derives its qualities and reputation from the place of origin, this reputation may become of high commercial value, and therefore be subject to suffer of IPR infringements by means of misuse, counterfeit, passing off or similar actions that may derive in the dilution of the reputation and good name of the product. This is why it is generally recognized that these signs need to be protected. GIs are given different names such as appellations of origin, designations of origin, origin signs, etc., in different national laws.

GIS Protection: International Treaties and Agreements Related with Intellectual Property Rights of Geographical Indications, and its Protection at National Level

As we learned in the first section of this chapter, there are different international agreements and treaties for IPR protection and in specific, we can name the following international instruments of the World Intellectual Property Organization (WIPO) and the World Trade Organization (WTO), related with Geographical Indications protection which is in force today:

- ☆ The 1883 Paris Convention on the Protection of Industrial Property;
- ☆ The 1981 Madrid Agreement for the Repression of False or Deceptive Indications of Source;

- ☆ The 1891 Madrid Agreement and the Protocol on the International Registration of Trademarks;
- ☆ The Lisbon Agreement for Protection of Appellations of Origin and their International Registration of 1958,2, and
- ☆ The WTO Agreement on Trade-Related Aspects of Intellectual Property Rights (Articles 22-24).

Mexico is part of the Paris Convention and of the Lisbon Agreement as well as of the TRIPS, but is not part of the 1981 Madrid Agreement for the Repression of False or Deceptive Indications of Source, nor is it a contracting party of the 1981 Madrid Agreement and the Protocol on the International Registration of Trademarks. In this sense, Mexico has joined the treaties related with GI´s more specifically and not the ones related with trademarks as private IP rights, which are today protected by the Mexican individuals in a registration basis per country or well through diverse other modalities of access to this protocol or to other regional systems available.

Returning to GIs protection, also national laws protect them. According to the type of law of each country, we will find that some countries protect them through a specific Law created for the protection of GI´s and others will do it through the trademark laws system, in the form of collective of certification marks, or in by means of laws against unfair competition, consumer protection laws or laws made specifically for the protection and recognition of individual GIs.

Recognition of a GI Protection

The protection of Geographical Indications as very simply stated in the study of O´Connor, can be obtained through the following systems:

- ☆ Via a Sui generis system in the countries or regions that count with a law that considers geographical indications as a specific, separate intellectual property right and/or
- ☆ Via Trademark regimes, found in the countries or regions where geographical indications are not considered as a separate, independent intellectual property right.

  In some systems names and signs that are protected as GIs, can also be registered as trademarks (individual, collective, or certification, where available).
- ☆ O´Connor study further explains in the same line of ideas, how the observance, making the legal protection effective is done in practice, which is at the end the most important for the right holders of the GI. Generally speaking, there are two main approaches: *ex officio* and *ex parte* systems:
- ☆ *Ex officio* this protection is offered by the countries that have a law for protecting GIs, the authorities will then be in charge and have the responsibility for policing the GI protection.
- ☆ *Ex parte* protection is the protection granted by the government once the interested party asks for it. Be it the producers, the regulating body, or other. In this kind of protection we may find that the interested party will

finance the protection activities that include registration expenses, suing in court, and all the administrative expenses related with keeping the protection acquired.

Legal Framework for the Geographical Indications in Mexico

In the case of Mexico, the country has an *ex officio* system for GI´s protection, which for the Mexican industrial property law is known as appellation of origin (AO).

In Mexico there is the possibility to protect a product as an appellation of origin through the Declaration made by the government, per the request of the interested party, once it proves a series of requisites of the existence of the AO. This declaration will then be the basis for the *ex officio* protection.

Also, there is the possibility of registering collective marks for products that have a particular characteristic and want to be protected by the interested group of producers, through the appointment of some Rules to control the use of the collective mark in order to maintain its good reputation. In the case of an infringement, the interested party may start an administrative procedure at the IMPI and go to court for their private rights being infringed just as with the registered trademarks. The Certification marks in Mexico are not yet available.

The Industrial Property Law (IPL from now on) of Mexico establishes in its Article 156 that "An appellation of origin shall be understood to be the name of a geographical region of the country that is used to designate a product originating therein of which its qualities or characteristics are due exclusively to the geographical environment, including both natural and human factors."

The Right Holder of the Appellations of Origin of Mexico is the Mexican State, as is written in the Industrial Property Law in its Article 167, therefore it is the State through its Governmental Agencies the one in charge of managing the recognition and the authorization of use of any appellation of origin of Mexico.

As I mentioned above, the scope of protection of the appellations of origin in Mexico is a Sui generis protection, all Mexican appellations of origin are protected Ex officio. Like for the other Intellectual and Industrial Property Rights defence and observance, many agencies are involved in the protection of the Appellations of Origin, as we saw above, the ministries of Economy, Health, Agriculture, Customs, Tax and Revenue and the Police, are all involved in some part of the protection of the Appellations of Origin of Mexico.

The IPL of Mexico offers a high level of protection to registered AOs: protection against unfair competition as well as against any use susceptible to cause error or confusion on the geographic origin of products. This protection covers the use of the GIs with expressions such as: "class", "type", "style", "imitation" or other analogous ones.

The ex officio protection applies not only to Mexican AOs today, but also to the European Union Spirits with GI that are part of the bilateral agreement concerning the mutual recognition and protection of designation of spirits, signed in May 1997 between the European Union and México.

The protection is to be granted by the Mexican government even where the true origin of the spirit drink is indicated or the designation is used in translation or accompanied by terms such as 'kind', 'type', 'style', 'way', 'imitation', 'method' or other analogous expressions, including graphic symbols which may lead to confusion (Article 5 of the Agreement Mexico UE-1997).

Foreign appellations of origin can be registered by the national authorities of foreign countries or by persons with legal interest provided that it is established in a Treaty of which Mexico is party.

The Mexican Institute of Intellectual Property (IMPI) can grant a declaration of recognition to the foreign appellations of origin. With such recognition, future trademarks in Mexico cannot be registered with the name of the appellation of origin.

There are three ways to obtain this recognition:

- ✰ Via the Lisbon Agreement of Appellations of Origin, as Mexico recognizes the appellations of origin of countries that are parties to this Agreement;
- ✰ Via notification sent by the WIPO;
- ✰ Through the request of a country or person with legal interest in the recognition by the IMPI.

There is no opposition procedure. A decision to reject registration can be appealed before the Mexican Federal Superior Court.

The main requirements for the recognition of an appellation of origin in Mexico are:

1. That the name of the appellation of origin is not registered in Mexico;
2. The certification of the country of origin certifying that the appellation of origin is registered there.

GIs are protected for an indefinite period of time (as long as the conditions for GI protection are respected).

Appellations of Origin of Mexico that are Protected Today and the Management of these Appellations of Origin

It was not until the year 1973 that the appellation of origin protection was recognized in the Industrial Property Law and this was done through a Decree made by the President of Mexico, Luis Echeverría in which the 1942 IPL was modified including the AO distinctive sign modality, and the story tells that this action was very much driven by the Tequila industrials that were very interested in getting this distinction for their already valued product.

Today there are 13 Protected Appellations of Origin in Mexico, 5 are distilled spirits, 5 are agro products and 3 are handicrafts made with specific raw materials coming from a specific region with particular traditions.

Given that the right holder of the Appellations of Origin in Mexico is the Mexican State, it is the State, in specific the Mexican Institute of Industrial Property, the Entity in charge of authorizing the use of the AO for the producers that can accomplish the requirements stated in the IP Law, such as being related directly with the production of the good in the delimited region stated in the Declaration of Protection, as well as proving that they are in compliance with the Official Standard for that AO.

The Official Standards known in Mexico as NOM (Norma Official Mexicana) are a fundamental part of the protection system for the appellations of origin in Mexico.

Each Declaration of an Appellation of Origin states the Territory of production or harvest of the produce, and specifies the Official Standard with which that AO will be regulated, all the producers that want to use the name of the AO will need to comply with it in order to obtain and to keep the authorization of use of the AO, granted by the IMPI.

The Official Standards are technical regulations and not laws, they are created by the General Direction of Standards together with the interested parties (the whole productive chain, producers, packers, marketers and government).The aim of the Official Standards is to determine in detail the specific requirements per product, its raw materials, and its processes with the aim of taking care of the peoples´ security and of protecting the human health, the animals and the plants, as well as the environment in general.

The official Standards are compulsory as stated in the Federal Law about Metrology and Standardization of Mexico, and as such their compliance must be verified.

In 1992 a modification in the Federal Law of Metrology and Standardization, made it possible for private organisms to verify the compliance of the Official Standards and with this the conformity assessment organisms or regulatory councils, came to life.

The Regulatory Councils need to be accredited in order to operate, by the Mexican Entity of Accreditations (MEA) after complying with the specifications that include proving to count with the specialized staff for the subject, with the technical capacity, material infrastructure, the quality management manuals, that will guaranty the development of their functions, which may include the verification unity, certification organism, and laboratory of tests. Finally the MEA will make and *in situ* audit in order to review the work in practice of the organism and see it functioning according to the manuals.

The regulatory councils for the AOs in Mexico are subject to the approval of the General Direction of Standards (Dirección General de Normas, DGN) which is the Governmental Agency in charge of ensuring that the Official Standards are respected.

Not all the AO count with regulatory councils to verify the compliance of their Official Standards however it is an option that has proven positive for the quality assurance in at least some of the most known Mexican AOs. These regulatory councils will verify the complete process from raw materials to packaging of the product getting finally to the certification of the product that complied.

The AOs that do not have a regulatory council are subject to the verification of the Government, since the Official Standards compliance are compulsory for every user of an AO; however, this verification is not totally devoted to their productive chain, since the government needs to cover a wide variety of activities, thus there is a possibility for the producers of deviating in the production specifications for quality and security assurance of some of the products, and that then reflects in the product success.

In the following section, I will point out a brief list of the AOs recognized in Mexico in chronological order up to date, as well as point out the ones that count with regulatory councils.

Tequila was the first Appellation of Origin to be recognized by the Mexican Government in the year 1974 when a Decree was published granting Tequila the protection as an appellation of origin according to the Industrial Property Law. It was until 1977 that the Declaration of the Protection for the Appellation of Origin Tequila was published. The 5 States where Tequila may be produced are Jalisco, Nayarit, Tamaulipas, Michoacan and Guanajuato. In year 1994 the Tequila Regulatory Council was accredited.

Mezcal, was protected by the Mexican Government in year 1994, it is also a distilled spirit made from a different variety of agave that is found in the States of Guerrero, Oaxaca, Durango, Zacatecas and San Luis Potosí. In year 2003 the Mexican Regulatory Council for the Quality of Mezcal (COMERCAM) was accredited.

Olinalá was declared protected in the year 1994 and is a handicraft made out of a very special aromatic wood in the locality of Olinalá, Guerrero. This AO has not yet and Official Standard therefore there is not yet a regulatory council to verify and certify its compliance.

Talavera, was declared protected in 1995 and is a traditional handicraft that has its origin in the colonization and became very specialized and specific to the zone of Puebla and Tlaxcala States in the Center of Mexico. The Consejo Regulador de la Talavera AC, was accredited in 1999, but in 2005 its accreditation was canceled and is in its process for recovering it.

Bacanora, was protected in year 2000.This product is a distilled spirit made out from agave whose production is limited in the State of Sonora in the north of Mexico. They do not count with a regulatory council but are in the process of consolidating one as most other AO of Mexico, they have the Consejo Sonorense de la Regulación del Bacanora and the Sonora Government and 15 Bacanora producers are working together to get the requirements to become accredited and approved.

Café Veracruz, was protected in year 2000 and is a coffee from Veracruz, the State bordering the Gulf of Mexico. The Consejo Regulador del Café de Veracruz CRCV, was accredited in 2003 in order to verify and certify the compliance of the Veracruz Coffee Standard.

Ámbar de Chiapas, was protected in year 2000 and is a special kind of amber, only obtained in this State of the South East of Mexico. It does not count yet with a regulatory council but the Government of the State of Chiapas is working together with the council in order for it to get accredited shortly, the government plans to help the council with consultancy.

Sotol, was protected in year 2002, it is another distilled produced from a different variety of plant very close looking to the agave called ***Dasylirion***, it can be produced in the States of Chihuahua, Coahuila and Durango in the North and North West of Mexico. The Sotol Regulatory Council is not created yet, it is in process of creation in the body of a civil association and has gotten consultancy from the Tequila Regulatory Council with the steps needed to succeed in this.

Charanda was declared protected in 2003 and is a cane sugar distilled spirit made in the State of Michoacán. 1985 the Association of the Charanda producers was created and they are working in the creation of their regulatory council.

Mango Ataulfo del Soconusco de Chiapas was protected in 2003 and is a fruit coming from one specific region of the southeast state of Chiapas. The Official Standard for this AO has not yet been created and there seems to be a lack of agreements amongst the State Government and the local producers of Mango in order to succeed in the creation of the Official Standard and with this be able to create the Regulatory Council to verify and certify the compliance of that Standard.

Café de Chiapas, protected in year 2003 is a special produce also from the south eastern state of Chiapas which gives its names to this coffee. In the same year the Chiapas Coffee Regulatory Council was created as a civil association with the support of the State Government of Chiapas but it is not yet accredited, therefore, it cannot certify any product as being part of the AO yet, but the Government is working with them in order to get accredited.

Chile Habanero de la Península de Yucatan, was declared protected in 2008, but it had to be changed because there was a controversy amongst the peninsula of Yucatan states since the protected territory included only the Yucatán region and not Campeche and Quintana Roo which are also States with Chile Habanero production. The actual version of its declaration was then issued in 2010. Its official standard is not yet created therefore they do not have a regulatory council to verify and certify it.

Vainilla de Papantla, is a vanilla from the State of Veracruz and Puebla from the specific location of Papantla, and it was declared protected by the IMPI in the year 2009. Its official standard is not yet created therefore they neither have yet a regulatory council in place.

In the case of the Mexican Appellations of Origin we can find that not always reality matches the theory which in this case would be the IP Law of Mexico, where it is written what should be recognized as an appellation of origin. In realitywe find out that there are three different types of recognized AOs in Mexico:

1. AOs that follow the IP Law requirement, that is they do have the same name as the region or geographical zone of the country, such as: Olinalá, Tequila, Charanda and Bacanora (pg 76 DO mexico)
2. AOs that use a generic name plus the name of a geographical zone, such as: Café Veracruz (Veracruz coffee), Café de Chiapas (coffee from Chiapas), Mango Ataúlfo del Soconusco de Chiapas (Mango from the Soconusco region of Chiapas State) and Ambar de Chiapas (Amber of Chiapas).

3. AOs that are named after another noun different from that of its origin, as Mezcal which is the name the people of Mexico give to the agave plant from Oaxaca; or Talavera: which is a locality from the province of Badajoz, Spain, and is not a place from Mexico but the handicraft with AO from Mexico took the name from that place. Finally Sotol is the name of the plant that originates in Chihuahua, Coahuila and Durango with which the beverage is made. All these products were recognized as AOs because the locals started naming them like that traditionally until each of them got an important reputation that allowed them to obtain the protection as AO from the Government.

In the Framework of the Lisbon Agreement the following Mexican Appellations of Origin are part of the International Registration and are therefore protected from becoming generic: Tequila, Mezcal, Olinalá, Talavera, Bacanora, Café Veracruz, Ámbar de Chiapas, Sotol, Mango Ataulfo del Soconusco de Chiapas, Charanda y Café Chiapas.

From the analysis of the information of the AO that Mexico has protected, we can see that the GI and AO recognition is a fairly new IP protection modality for the country: the oldest recognition of an AO is from 1974, and the last one was just recognized in year 2009.

It is also interesting to see that, even though it is compulsory for an AO to count with an official standard, and that the compliance of it would be the basis for the authorization by the Mexican State to use the AO; many of the Mexican AOs do not count with an official standard yet, in fact 4 out of 13 do not have one, and 9 of 13 do have an official standard to comply with but only 3 out of 13 do count with an accredited Regulatory Council in order to verify and certify this compliance. Thus, the Mexican AOs are being born, yes, but many of them are not having enough supervision in their upbringing.

The IPR system regarding AOs in Mexico needs to be improved much in order to ensure the consumer that he is really getting what he is expecting, as we see from the lack of a proper control over all the AOs that have been recognized and protected.

What is important to learn from this information I believe, is that we are missing an opportunity as Mexicans for getting all the benefits that controlled AOs may give, not only for the producers by getting their product sold better, but for the consumers around the world who would enjoy the highest quality certified products, had all the AOs an official standard and a regulatory council devoted to their certification. As it can be very well synthesized in the words of JG Murillo:

"Let us say it in another way, when a product is associated with an appellation of origin, exists behind it a technical standard of production, respected by the whole productive chain, and a regulatory council that verifies and certifies the superior quality of the product, this validation gives this distinctive sign an important value as a penetration and acceptance mechanism in the market, which fosters the preference of the consumers".

In the following section I will focus on the Tequila case, as to show how it is possible to take advantage of the mechanisms available today in order to make the most of an AO, having of course many possible areas of improvement still to consider.

The Appellation of Origin of Tequila (AOT) and its Regulatory Council

Name, History and the Declaration of Protection of Tequila

First of all it is important to locate the geographical situation of Tequila and its meaning for the people that lived there in pre-colonial times.

"The word tequila is of Nahuatl origin and has different meanings, the most probable is "place of wild herbs", but some say it means "place where one cuts" or "place where one works". Tequila is the name of a town of pre-colonial origin, located at 1218 above sea level, and at less than 60 kilometres at the northwest of Guadalajara"

In year 1795 is when the first concession for producing a mezcal wine of Tequila was granted by the spanish Crown, since before that it was prohibited by the Spanish Monarchy to its Colonies to produce any local wine or spirit.

Since 1943 some lobbying was being done by the industrials of the region, in order to protect the name "Tequila" to obtain the exclusivity of its use. The arguments being centered in the large story associating the industry with the region with that name (region, town and hill are all called "Tequila") with this alcoholic drink that has earned the name of "The National Drink".

It is in June, 1949 that the first Official Standard for the Quality of Tequila is published "DGN R-9-1949". This Standard as all the Official Standards in Mexico is reviewed and updated every 5 years.

As we learned in the previous section, the Appellation of Origin of Tequila was the first one to be recognized by the Government in 1974 by the Declaration for the Protection of the Appellation of Origin of Tequila (AOT). This declaration requires that the production of Tequila comply with the provisions set forth by the Official Mexican Standard of Quality for Tequila, which is established by the General Direction of Standards.

The Mexican Official Standard for Tequila in force today and since April 2006, is the NOM-006-SCFI-2005 Bebidas Alcohólicas- Tequila- Especificaciones", establishes the specific requirements that everyone from the productive chain Agave- Tequila should accomplish in order to bear the name Tequila. The Standard establishes the requirements for the agave supply, process of elaboration, bottling and marketing practices linked to the Tequila spirit, and any Tequila producer should prove its strict compliance with this Standard.

Tequila is an alcoholic drink made from the fermentation and distillation of the extracted juices of the *Agave tequilana* Weber blue variety. The Agave can only be planted and harvested in Mexico within the limits of a specific region, namely, the Region of the Appellation of Origin of Tequila.

The Territory. The Mexican States selected to compose the region of the Appellation of Origin of Tequila - according to their climate, altitude, soil and use of traditional techniques of production- are the state of Jalisco in its entirety (125 municipalities) and some municipalities of the states of Guanajuato (7), Michoacán (30), Nayarit (8), and Tamaulipas (11), adding up to a total of 181 municipalities in 5 different Mexican States.

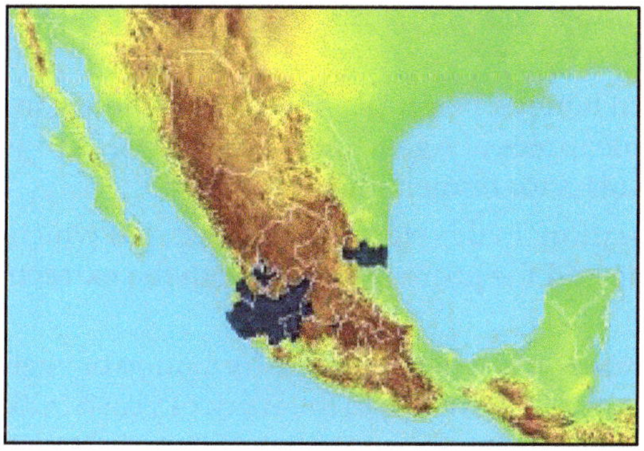

Figure 11.2: Map of Mexico and the Territory of the Appellation of Origin Tequila as stated in the Declaration of Protection for the Appellation of Origin of Tequila, 1974.

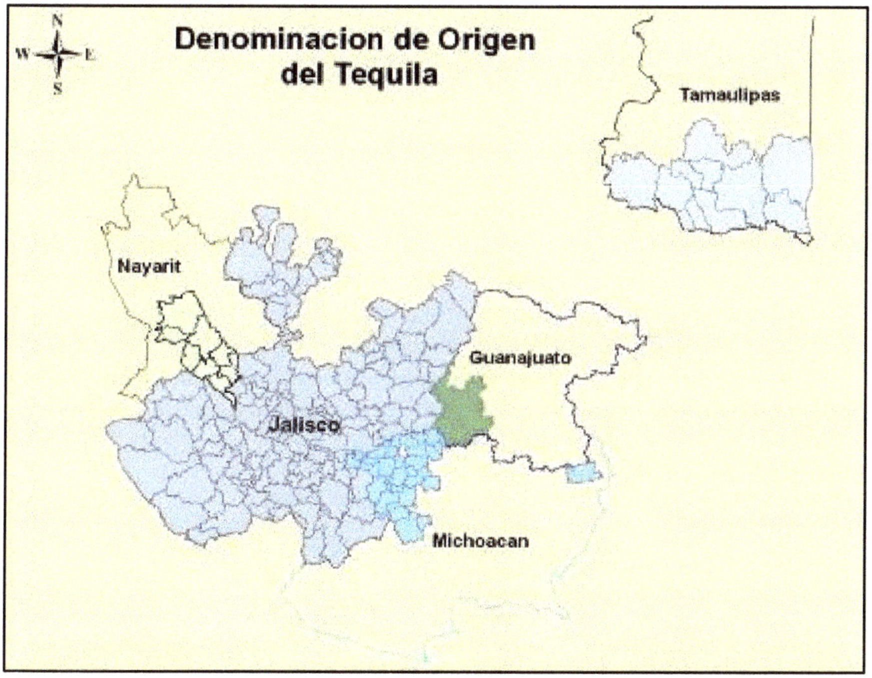

Figure 11.3: Zoom in to the Appellation of Origin Territory and its 181 Municipalities.

The Tequila Process

1. **The "Jima"** consists in the cutting off the leaves of the plant *Agave tequilana* Weber blue variety down to its base, to keep only the head or heart of agave, which is the raw material of Tequila.
2. **Cooking or Hydrolysis**. Is the process of transformation of the complex starches of the Agave into simple sugars, which are easily fermented and converted into alcohol, this is done through heat or other methods like hydrolysis.
3. **Extraction of honeys**: Before or after cooking, the sugars or starches obtained from the pineapples of Agave must be separated from the fibre in order to use these sugars for fermentation.
4. **The formulation**: In this step it will be decided which category of Tequila will be done and the juices obtained from the extraction are prepared for fermentation.
5. **The fermentation** is where the transformation of vegetal sugars into ethyl alcohol is carried out thanks to the work of yeasts. Other components will also be formed during this step that will contribute to the final sensorial attributes of Tequila.
6. **The distillation** is the process by which different components of the fermented mixture are separated, by means of heat and pressure, obtaining purified alcohol and letting go the undesirable residues. All Tequila must be distilled at least twice.
7. **Aging or Maduration** is the slow transformation of Tequila that lets it acquire additional sensorial characteristics like odour, colour and texture by physical chemical processes that in a natural way have place during its storage in recipients or barrels of wood oak or holm oak. The Extra Aged and Ultra Aged Tequilas must be matured in wooden barrels not larger than 600 litres.
8. **Bottling** is the action of pouring the Tequila in the recipients that will contain it in order to keep it and protect its physical and chemical stability and market it. Tequila must be bottled in new recipients, allowed by the sanitary authorities. The range of alcoholic content allowed by the Standard of Tequila is from 35 to 55 alcohol percentage by volume.

Classification of Tequila

Tequila Categories

Tequila 100 per cent agave. It refers to Tequila that has been produced from only the fermented and distilled juices of the *Agave tequilana* Weber blue variety. **Bottled exclusively at the AOT.**

Tequila. Made with a mix of sugars, where at least 51 per cent of these sugars must come from Blue *Agave tequilana* Weber, and the remaining 49 per cent is made of sugars from other sources. **It can be bottled abroad.**

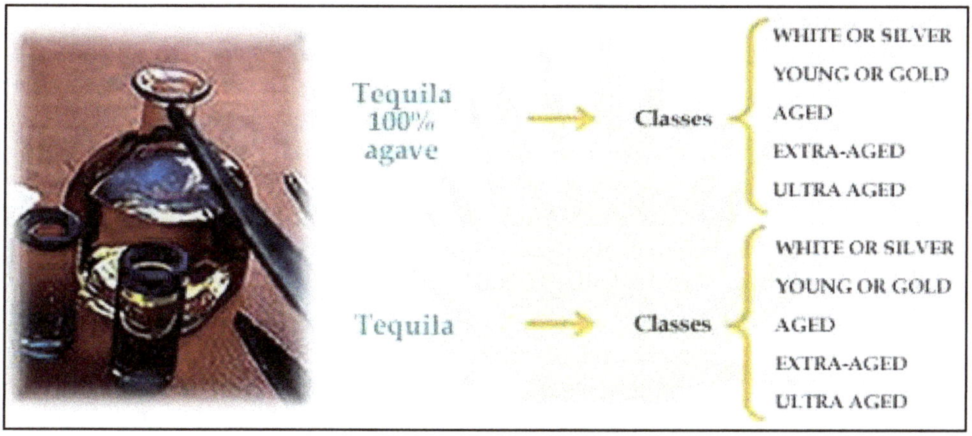

Tequila Classes: the same for each of the 2 categories

White or Silver Tequila

Is basically a crystal clear product that is not aged and can be bottled immediately after the second distillation by adding distilled water for obtaining the commercial alcohol volume.

Young or Gold Tequila

Is non-aged Tequila that acquires its yellowish colour through authorized additives, such as caramel colouring, or that results of a mix of white Tequilas with aged Tequilas.

Aged Tequila- Reposado

It is the product that has matured a minimum of 2 months and a maximum of 12 months in white oak barrels, giving the Tequila the colour and flavour of wood.

Extra-aged Tequila- Anejo

When the Tequila is aged in barrels with a maximum capacity of 600 litres for at least a year, it may belong to this category. The colour, aroma and flavour of the wood are more pronounced in this type of product.

Ultra-aged Tequila- Extra Anejo

Is the product with an aging process that lasts at least three years in direct contact with the wood of oak or Encino oak (holm or holm oak) in containers with a maximum capacity of 600 litres.

Tequila 100 per cent Agave and Tequila with Flavour in any of its Classes

The Tequila Standard NOM-006-SCFI-2005 allows the addition of sweeteners, colouring, aromatizes and/or flavourings to Tequila in certain amounts, in order to generate products with various colours and flavours and widen in this way, the options for the Tequila consumer.

The Tequila Regulatory Council

The Tequila Regulatory Council (CRT) is an institution that was created as a pillar for the protection of the Appellation of Origin Tequila which is established in 1994. The CRT is a private organization in charge of verifying that all Tequila produced complies with the provisions set forth in the Official Mexican Standard for this product. After a satisfactory verification, that includes the permanent in site inspection in every part of the process, the CRT issues the certificate that proves the product is authentic Tequila.

In order to do its work the CRT is approved by the General Direction of Standards of the Ministry of Economy of Mexico and is accredited as a Unit of Verification, as an Organism of Certification and as a Tests Laboratory by the Mexican Entity of Accreditations (ema). The CRT has a system of Quality Assurance that guarantees the trustworthiness of its services. From June 1999, CRT obtained the ISO-9002 certification, which means it has a permanent commitment to improve its processes with a culture focused towards quality.

The CRT is a strong defender of Tequila. With international offices in Washington D.C., U.S.A.; Geneva, Switzerland; Madrid, Spain; and Shanghai, China, it is constantly monitoring the fulfilment of the mandatory Standard for Tequila (NOM-006-SCFI-2005) and that the Appellation of Origin of Tequila is respected all over the world by ensuring that only authorized and certified products bear the name Tequila.

With the aim to guarantee the representation of the whole Tequila sector and the transparency in its operation, the CRT holds these 4 sections in its Board: the tequila producers, the agave producers, the bottlers and distributors, and the governmental representatives related with this industry. All of them in a balanced and participating way help guarantee the credibility and participation of the whole productive chain in the Councils decisions.

The main objectives of the Consejo Regulador del Tequila (Tequila Regulatory Council) are:

1. Ensure, through the verification process, the fulfilment of the "Norma Official Mexicana- NOM (Official Mexican Standard) for Tequila.
2. Safeguard the Appellation of Origin Tequila in Mexico and around the World.
3. Guarantee the authenticity of the Tequila that the consumer acquires
4. Issue timely and truthful information useful for the Agave- Tequila productive chain.

Additional to the certification and verification duties, the Tequila Regulatory Council does some other activities, which are vital for the fight against the disloyal competence present in the market, nationwide and internationally, as we will see in the next section where I will focus on the infringements, the prosecution by the authorities, and the measures taken to face those infringements by CRT.

Finally about the CRTs work, it is important to say that through its interdisciplinary staff it has established a plan of strategic research related to the agave production, for helping out the primary sector improve their practices in the field as well as related to the industry, in order to help them better manage different challenges from the process to the management of waste waters, or related to technical barriers to trade imposed by these countries in which the Tequila Regulatory Council works together with the Government as Technical support to provide all the information necessary to dilute those barriers. All these additional activities are always directed towards the protection of this Mexican symbol, and also towards creating a culture of tequila in the consumer.

The CRT manages two programs for the promotion and protection of Tequila from a different perspective not related with the verification, certification nor the prosecution of the fake product, which have proven to further enhance the knowledge and respect for Tequila.

The Award T program (Distintivo T- DT for its Spanish acronym) is a training course the CRT gives to the establishments that sell or distribute Tequila, so their staff is totally involved in the Tequila Culture, they learn about the Mexican Official Standard, the Agave Plant, the Protection of the Appellation of Origin, the Production Process, and how to Taste the product passing by each of its classes. The establishments compromise by it to only buy certified products and to destroy the empty bottles to avoid re filling of bottles. Up to date the CRT has given 195 Award T´s worldwide, 170 in Mexico and 25 abroad.

The Tequila Trail is a Touristic Regional Development Project that is being sponsored by the Interamerican Development Bank and a Foundation of the Region in order to support 400 middle and small enterprises so they improve their services and touristic products in order to attract more tourism to the region, and foster economic development, all this in the framework of the declaration as a World Heritage site by UNESCO of the Agavero Scenary in 2006.

Scheme of Primary Protection of Tequila

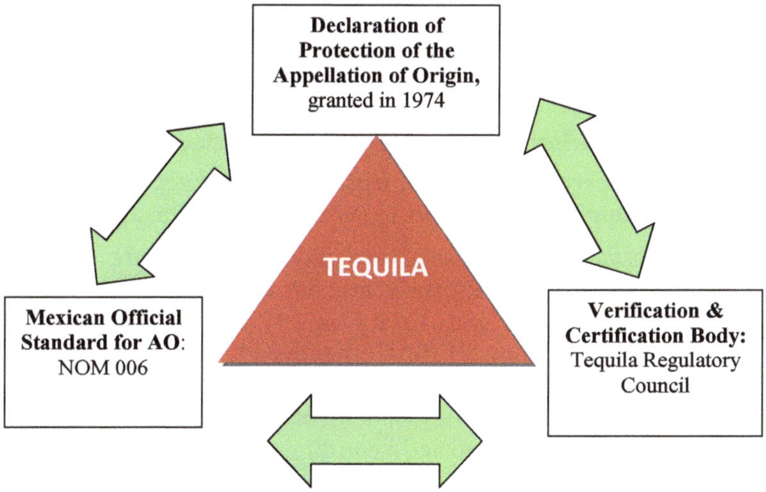

Infringements Against the IPR´s of Tequila Nationally and Internationally

The products that get to have a good reputation will most of the times see their property rights violated, this is the case of Tequila. The infringements we find against the IPR of Tequila today, nationally and internationally are the following:

Counterfeiting: Factories or bottlers producing or bottling other beverages and naming them Tequila without compliance of any standard, usually is cane sugar alcohol.

Fraud: Refilling of empty Tequila bottles being then sold to the public as Tequila or use of trademarks that include the word tequila in them in order to confuse the consumer about the true origin of the product.

Unfair competition: through products that are using the name Tequila such as beers, liquors, even foods that claim to contain tequila but are not linked to the industry in any way.

Use of the name without a license from the right holder: some producers of beverages made of agave claim they sell tequila, without being certified hence have not gotten the authorization from the IMPI to use the name Tequila.

Passing off: here are included the products which through the association of images related with Tequila -such as agaves, haciendas, or other Mexican icons-, try to confuse the consumer without using the name tequila itself. Here could be included the agave distilled spirits that are sold to the public as Tequila.

Smuggling: Some agave distilled spirits (which do not have a standard or verification control at all) or even cane sugar alcohol have been found to be exported as such and then be imported into the other country as Tequila.

All these infringements lead to a series of consequences such as potential health risks due to the lack of control of those products, the damage of the reputation of Tequila, the possible dilution of its name and the negative impacts on the formally established tequila industry, since the producers that are complying with the standard will be selling much less.

Prosecution and Protection Strategies by the Government and by the Tequila Regulatory Council

Monitoring of the Market and Prosecution in Mexico

In Mexico as we learned in the prior pages, the IPR protection is done by different agencies in a coordinated manner and so is the case with the appellations of origin.

For the Tequila case we have some vital agencies in charge of the prosecution of the infringements against its IPRs, these are: the Attorney General of the Republic (PGR), the Mexican Institute of Industrial Property (IMPI), the General Direction of Standards (DGN), The Agency for the Consumer Protection (PROFECO), and the Federal Agency Against Health Risks (COFEPRIS). Since the infringement against an AO is a federal crime it is the Attorney General of the Republic the one to prosecute it, however they will need in order to act, a complaint filed by the Mexican Institute of

Industrial Property, being this institution the head authority regarding the AO. The PROFECO verifies and monitors the compliance of the Mexican Official standards in relation with labelling of alcoholic beverages and they have found that the practices above mentioned are a real social problem against the national consumer since he is being deceived constantly.

The Tequila Regulatory Council conducts a regular program of verification in the retail stores along the Mexican Republic, which consists in monitoring what is sold, checking its labelling and trademarks as to make sure they comply and that only certified product is sold. When the CRT finds out a product that is not certified then it opens an investigation file which is handed in to the Consumer Protection Agency in order for them to act since the CRT does not have as a regulatory council any attribution for punishing, sanctioning or similar. The CRT may only report and the authority may proceed.

The problem with this prosecution system is that in Mexico the infringements against an AO are not typified as felonies but only as minor offense, and therefore they will only receive administrative sanctions such as fines, closing of their facilities and seizures; in order for this minor offense to become a felony or major crime it is to repeat the same administrative fault or minor offense once this minor offense has been totally proven, which leaves the AO offenders the door open to keep infringing against the AO´s rights.

In Mexico today the Industrial Property Law considers it as a felony to counterfeit or misuse a commercial trademark, while it does not consider so the use of the AO by someone with no right to use it, until he has done it for the second time. This request has made it almost impossible to get a criminal sanction for any AO offender, since any offender may create another enterprise to start again the infraction against an AO without falling in the category of second time offender.

The lack of exemplary actions against the AO offenders due to a legal constraint made by the law itself, leaves the AO practically unprotected against the free ride pirates, and leads to a dependence on the trademark right holders, authorized users of the AO, in order to protect the Tequila name, leaving it up to them to sue against the offender of their trademark when it does happen, or not, which may be very dangerous for the Tequila name, since many individuals are not interested in fighting so the infringements keep happening and the dilution and bad image of tequila may be growing.

Monitoring, Prosecution and Preventive Measures for Protection Abroad

Infringements against Tequila IPRs are found either in countries where Tequila is a well-known product or in countries where they feel they are so far away that no one will ever notice; however Tequila is today exported to more than 120 countries, therefore the Tequila distributors are literally everywhere, and it is through them that many of the reports about counterfeit products reaches the CRT.

The most common practices of infringements found in the foreign market today are products that claim to be Tequila but are not linked to the Industry in any way, some of them even claim to be produced in other regions of the world with the Mexican recipe or just to be made with Mexican ingredients, which totally denotes the lack of knowledge of the AO Tequila.

CRT counts on different resources to monitor the international market for counterfeit products, one of them is through its international offices that carry out weekly in site inspections to the retail stores in order to verify all the Tequila sold there is certified, also they search in the world wide web for counterfeit products, any case that is found will be investigated in documents and records and if its sample is available it will be analyzed.

Even certified products are subject to be analyzed for the record and for making sure the chromatographic profile of the analysis has not been modified, in order to review that, the verification unit of CRT compares the test results per batch that are kept in the CRTs electronic archives per factory if they do not match the product was adultered and the authorized producer is advised so he can reply before being sanctioned by any national authority.

The tests results of presumable counterfeit products are taken to the Verification Body of CRT which will be able to issue a technical report where it would be stated that the parameters do not meet the required by the Mexican Official Standard for Tequila and this report is sent to the local authorities of the country where the product was found or to the Mexican Embassy of the locality in case CRT is not acquainted with the local authority yet. Through this technical report the local authority may then ban from the local market the counterfeit product, many times in the basis of their consumers' protection.

In order to prevent further infringements and better protect the Appellation of Origin Tequila IPRs, CRT has launched since 2003 but most aggressively since 2008, a Shield Project for the AOT. Through this, the CRT aims to protect most of Tequilas exports markets by the end of year 2012, through the registration of Tequila as GI, AO, Certification Mark or Collective Mark, as applicable. So far CRT counts with 7 granted protections around the Globe, 5 Registers as Appellation of Origin in Uruguay, Costa Rica, Ecuador, Honduras, Panamá; 2 Certification Marks one in China in both characters latin and Chinese, and another in Canadá granted to the Mexican Government since 1992; and is in process of getting the application for 8 more countries.

Another important protection action is the fact that The Tequila Regulatory Council is member and is actually holding the Presidency of ORIGIN the International Network of the Geographical Indications. This NGO was founded in year 2003 and the CRT has been part of it from the beginning, it gathers more than 2 million producers from 150 organizations of 40 countries around the world, and represents an important lobbying agency for supporting the GI´s aims and needs since it has its office in Geneva and is taking part in the major WTO and TRIPS negotiations regarding GI´s. Through ORIGIN the CRT has been able to speak up in defence of other GI´s, has networked with many other GI´s, and become more aware of the challenges GI´s are facing around the world.

Law Initiatives to Improve the IPR System of Mexico Related to Appellations of Origin

Due to the above mentioned, there are today two Law Initiatives aiming to improve the IPRs protection of the AO in Mexico; one is at the Senate of the Republic and the other is at the Jalisco State Congress.

The first was presented in September 2008, and was an Initiative with Decree Project to add article 224 bis to the Industrial Property Law and modify the fraction VII of article 194 of the Federal Code of Criminal Procedure. The proposal is to sanction whoever uses an appellation of origin without being authorized, or those who, having the authorization, use it wrongfully; it would also sanction whoever that provide with the raw material for a product that uses the appellation of origin wrongfully. In the other hand, this initiative defines the term "use" and establishes that in order to determine if an appellation of origin is being wrongfully used it shall require the technical opinion of the evaluation body (regulatory council).

The second was presented at the State Congress of Jalisco by the Deputy Héctor Álvarez Contreras, in March 24, 2010 with two initiatives: the first is to reform and add article 252 to the penal code of the State of Jalisco, so as to change the administrative sanctions for penal sanctions against the offenders that commit consumer fraude due to the non compliance of the Official Standards.

The second was an Initiative of Legislative Agreement asking the Federal Congress of Mexico to modify and add articles 6,211, 225 of the Industrial Property Law; articles 464 of the General Health Law and article 194 of the Federal Code of Criminal Procedure. With this changes it is looked for to Grant IMPI the power to denounce anyone that may be responsible for the production of adultered beverages at the Public Ministry (MP for its Spanish Acronym, who are theprosecutors of any felony), int his initiative they propose to change also the Health Law and the Penal Code establishing that the alcoholic beverages adulteration is a felony.

Conclusions and Recommendations for Improving the IPR´S Protection of the Appellation of Origin of Tequila

The Industrial and Intellectual Property system of Mexico is willing to give the framework for a real recognition and protection of the IP rights and that can be inferred from the signed agreements by the country; however it is evident that the specialty and specificity of the management of the Appellations of Origin is leaving a gap for many of the Mexican AOs to not gain all the reputation they could, since they are not being subject to an onsite verification or do not count with a Standard yet.

That is a legal problem that needs to be solved, in my opinion it should be really strict and get established that all the recognized AOs should count first with an Official standard because otherwise there is no certainty about the product or produce quality. Then a very important step would be that all AOs could count as well with a verification body that would ensure the compliance of that standard in a permanent basis, this would make a large difference in the quality control of every product and would really guarantee the permanence of the authenticity of the AO and an added value that the consumer would pay

In the other hand, regarding the prosecution system for the AO offenders, where an infringement against an AO is considered a minor crime, I do think the actual legal framework of Mexico is allowing that many take advantage of an asset they are not entitled to use and therefore the value of the AO may be diminished by the misuse, denigration or bad reputation they may generate.

Today with the actual lack of punishment to the AOs offenders, the producers of an AO are having their IP asset less protected in house than in many other countries of the world, where the use of an AO without authorization is considered a major crime, like it is in Spain, for example; therefore, I do consider it very necessary that the above mentioned law initiatives are approved by the Senate and by the State Congress.

Finally, I do think that Mexico as a country formed by its institutions has very much to develop regarding Appellations of Origin Protection still, since the country has been working in other IP aspects and is each time more involved in the international arena in subjects that include the fight against piracy and counterfeiting products, copyrights and other very specific subjects it is also very important that as a country it looks inside and sets up the tools to control directly the counterfeiting groups that are damaging the Appellations of Origin which are a public asset and therefore an invaluable one that should be well kept by all.

Acknowledgments

I want to thank the Tequila Regulatory Council as my employer and team that has given me the opportunity to learn about the amazing world of Appellations of Origin, and of IPR and in special I thank Mr. Ramón González Figueroa, Director General of CRT for his support and teachings of the AOs world. I want to thank the Mexican Institute of Industrial Property for the invaluable support they have always given me for learning and for achieving many of the objectives we have as CRT. I thank the Indian Government and the NAM S&T Centre, for giving me the opportunity to learn from them and share our experience with other countries. To the government of the state of Jalisco I thank for helping me out to be part of this training course together with the Federal Mexican Government.

Finally, I want to thank my dear parents, my family and friends for their encouraging words and actions that make me feel strong at every new challenge.

References

International Affairs Section of the Mexican Institute of Industrial Property Web Page, last updated November 2009. "International Legal Framework on Intellectual Property", Mexico. http://www.impi.gob.mx/wb/IMPI/marco_juridico_internacional_en_propiedad_intelect (accessed June 2010)

World Intellectual Property Organization Web Page. In About WIPO- Treaties Section; "WIPO Administered Treaties" http://www.wipo.int/treaties/en/(accessed June 2010)

International Affairs Section of the Mexican Institute of Industrial Property Web Page, last updated November 2009. "International Legal Framework on Intellectual Property", Mexico.

http://www.impi.gob.mx/wb/IMPI/marco_juridico_internacional_en_propiedad_intelect (accessed June 2010)

Presidency of the Republic of Mexico. Plan Nacional de Desarrollo (National Development Plan) http://pnd.calderon.presidencia.gob.mx/eje1.html (accessed in June, 2010)

International Affairs Direction. 2010. Mexican Institute of Industrial Property-IMPI-. Questionnaire prepared for the Regional Meeting of IP Offices of Latin America. Santiago de Chile, Chile.

Ministry of Economy. Sectorial Program by Sections 2007-2012 http://www.economia.gob.mx/img-sip/Derecha/Bannerssip/progsectorial-economia.pdf (accessed November, 2009)

Official Gazette June 12, 2009. Deputy Chamber of Mexico. http://www.google.com.mx/search?hl=es&source=hp&q=LEY+DE+CIENCIA+Y+TECNOLOG per cent C3 per cent 8DA&meta=&rlz=1R2RNTN_esMX348&aq= f&oq= (accessed November 2009)

International Affairs Direction. 2010. Mexican Institute of Industrial Property-IMPI. Questionnaire prepared for the Regional Meeting of IP Offices of Latin America. Santiago de Chile, Chile.

Amigo Castaneda J. IMPI Webpage. http://www.impi.gob.mx/wb/IMPI/quienes_somos (accessed June, 2010)

International Affairs Direction. 2010. Mexican Institute of Industrial Property-IMPI-. Questionnaire prepared for the Regional Meeting of IP Offices of Latin America. Santiago de Chile, Chile.

World Intellectual Property Organization Web Page. In WIPO-"Geographical Indications" Section: at: http://www.wipo.int/geo_indications/en/(accessed June 2010)

O´Connor and Company European Lawyers. Insight Consulting. Part I of the Guide "Geographical indications and TRIPs: 10 Years Later…A roadmap for EU GI holders to gain protection in other WTO Members". http://trade.ec.europa.eu/doclib/docs/2007/june/tradoc_135088.pdf (accessed in June 2010)

World Intellectual Property Organization Web Page. In WIPO-"Geographical Indications Treaties" Section: at: http://www.wipo.int/geo_indications/en/treaties.html

World Intellectual Property Organization Web Page. In WIPO-"Geographical Indications" Section: at: http://www.wipo.int/geo_indications/en/(accessed June 2010)

O´Connor and Company European Lawyers. report commissioned and financed by the Commission of the European Communities.2007. Part II of the Guide "Geographical indications and TRIPs: 10 Years Later…A roadmap for EU GI holders to gain protection in other WTO Members". http://trade.ec.europa.eu/doclib/docs/2007/june/tradoc_135089.pdf (accessed in June 2010)

Mexican Institute of Industrial Property Web Page, last updated October 2009. English version of the Industrial Property Law can be found at: http://www.impi.gob.mx/wb/impi_en/industrial_property_law (accessed June 2010)

BIS

Agreement between the European Community and the United Mexican States on the mutual recognition and protection of designations for spirit drinks Official Journal L 152, 11/06/1997 P. 0016 – 0026 http://eur-lex.europa.eu/LexUriServ/LexUriServ.do?uri=CELEX:21997A0611(01):EN:HTML (accessed June 2010)

García Murillo J.G.,2009. Las Denominaciones de Origen en Mexico, Consejos Reguladores y Eficacia Jurídica. (The Appellations of Origin in Mexico, Regulatory Councils and Legal Efficacy). Universidad de Guadalajara, Centr Universitario de Ciencias Sociales y Humanidades, pp.40 available at: http://148.202.18.157/sitios/catedrasnacionales/material/2010b/garcia/ldenom.pdf (acceded June 2010)

Ley Federal sobre Metrología y Normalización (Federal Law about Metrology and Standarization), referenced from García Murillo J.G.,2009. Las Denominaciones de Origen en Mexico, Consejos Reguladores y Eficacia Jurídica. (The Appellations of Origin in Mexico, Regulatory Councils and Legal Efficacy). Universidad de Guadalajara, Centro Universitario de Ciencias Sociales y Humanidades, pp.78-83

García Murillo J.G. 2009. Las Denominaciones de Origen en Mexico, Consejos Reguladores y Eficacia Jurídica. (The Appellations of Origin in Mexico, Regulatory Councils and Legal Efficacy). Universidad de Guadalajara, Centro Universitario de Ciencias Sociales y Humanidades 92- 94 available at: http://148.202.18.157/sitios/catedrasnacionales/material/2010b/garcia/ldenom.pdf (acceded June 2010)

BIS, pp 122-185

World Intellectual Property Organization Web Page. In About WIPO- Treaties Section; "WIPO Lisbo Express Search" per country, Mexico: available at: http://www.wipo.int/cgi-lis/guest/bool_srch5?ENG+17 (accessed June 2010)

García Murillo J.G.,2009. Las Denominaciones de Origen en Mexico, Consejos Reguladores y Eficacia Jurídica. (The Appellations of Origin in Mexico, Regulatory Councils and Legal Efficacy). Universidad de Guadalajara, Centr Universitario de Ciencias Sociales y Humanidades, pp.17. Available at: http://148.202.18.157/sitios/catedrasnacionales/material/2010b/garcia/ldenom.pdf (acceded June 2010)

MURIA, José María. 2001. Una bebida llamada tequila.(A drink called Tequila) 2nd ed. Ágata, México, 2001, p. 15.

Tequila Regulatory Council information from the archives, some may be found at. www.crt.org.mx

García Murillo J.G.,2009. Las Denominaciones de Origen en Mexico, Consejos Reguladores y Eficacia Jurídica. (The Appellations of Origin in Mexico, Regulatory Councils and Legal Efficacy). Universidad de Guadalajara, Centr Universitario de Ciencias Sociales y Humanidades, pp.41. Available at: http://148.202.18.157/sitios/catedrasnacionales/material/2010b/garcia/ldenom.pdf (acceded June 2010)

Tequila Regulatory Council information, at. www.crt.org.mx

Tequila Regulatory Council information at. www.crt.org.mx and from the Mexican Official Standard for Tequila NOM 006 available at: Web page of the Ministry of Economy: http://www.economia-noms.gob.mx, or at http://200.77.231.100/work/normas/noms/2006/006scfi2005.pdf (accessed June 2010)

Ministry of Economy web page: http://www.economia-noms.gob.mx, directly the NOM 006 may be found at http://200.77.231.100/work/normas/noms/2006/006scfi2005.pdf (accessed June 2010)

Mexican Institute of Industrial Property webpage, Declarations, available at: http://www.impi.gob.mx/wb/IMPI/declaracion_general_de_proteccion_de_la_denomina11 (accessed June 2010)

Mexican Official Standard of Tequila at the Ministry of Economy web page: http://www.economia-noms.gob.mx, directly the NOM 006 may be found at http://200.77.231.100/work/normas/noms/2006/006scfi2005.pdf (accessed June 2010) pp 7.

BIS pp 7-8

Consejo Regulador del Tequila. Distintivo T web page. Available at: http://www.distintivot.org.mx/(accessed 2010)

Tequila Trail Web page available at: www.rutadeltequila.com.mx (accessed June 2010)

From the Tequila Regulatory Council Archives. 2010. Information obtained from the staff in charge of the finished product verification program of CRT and from the International Affairs archives, typology of the cases of infringements against the AOT.

García Murillo J.G.,2009. Las Denominaciones de Origen en Mexico, Consejos Reguladores y Eficacia Jurídica. (The Appellations of Origin in Mexico, Regulatory Councils and Legal Efficacy). Universidad de Guadalajara, Centr Universitario de Ciencias Sociales y Humanidades, pp.80-81 available at: http://148.202.18.157/sitios/catedrasnacionales/material/2010b/garcia/ldenom.pdf (acceded June 2010)

Mexican Institute of Industrial Property Web Page, last updated October 2009. English version of the Industrial Property Law can be found at: http://www.impi.gob.mx/wb/impi_en/industrial_property_law (accessed June 2010)

International Affairs, Tequila Regulatory Council, from the archives. 2010. Typology of cases. http://www.origin-gi.com/(accessed June 2010)

Senator Ramiro Hernández García, Law Initiative Presented in September 4, 2008. Senate of the Republic of Mexico Webpage, available at: http://www.senado.gob.mx/juridica.php?ver=D&ano=3&tp=&sel=2&idd=985

Deputy Héctor Álvarez Contreras, Law Initiative Presented in March 24th, 2010. Congress of the State of Jalisco Webpage, search per subject and date of the presentation of the law initiative, available at: http://congresojal.gob.mx/Servicios/sistemas/SIPLE/siple.cfm (accessed June 2010)

Chapter 12

Intellectual Property System in Panama: Biodiversity and Traditional Knowledge

Catherina Caballero-George[1] and Leonardo Uribe[2]

[1]*Institute for Scientific Research and High Technology Services (INDICASAT-AIP), Bld. 219, City of Knowledge, Clayton, Panama, Republic of Panama*
E mail: ccaballero@indicasat.org.pa

[2]*Intellectual Property Office, Unit of Access to Genetic Resources, National Authority of the Environment, Bld. Plaza Albrook, Office #17, Panama, Republic of Panama*
E-mail: leonardo.uribe@anam.gob.pa

ABSTRACT

A major concern for developing countries is the fact that the intellectual property rights system tends to increase the costs of access to many products and technologies in areas of particular developmental importance like health, agriculture, education and information technologies. Like other developing countries, Panama has reacted protecting its genetic resources and traditional knowledge. Furthermore, Panama has followed the lead of countries like India, China and Brazil, in cataloguing its traditional knowledge and preparing inventories of its genetic resources, in order to check claims of invention against existing traditional knowledge.

Concerns such as software, which relates to conservation and information that relates to sustainable use of biodiversity may be the subject of copyright protection. In our particular case, much scientific information related to conservation and sustainable development is found in databases of restricted access. The government is pushing, through the National Secretariat of Science and Technology (SENACYT) to a more open access to this information, especially to governmental institutions focused on regulation and protection of the national biodiversity, as well as those of academic scope.

Indigenous knowledge and collective rights for indigenous groups have been covered in Panamanian law, however traditional knowledge based on genetic resources remains unprotected. In Panama, product patents have been granted on DNA sequences, genes, the products of genes and micro-organisms, as long as an application is involved. While, transgenic plants are protected through other mechanisms.

Panama concur that developing countries need a combination of increased international funding, specific initiatives to support research into areas that lack a lucrative market (neglected diseases like malaria, dengue, etc), and support for increased education, technology transfer and training within developing countries. The views on Patent system in Panama in relevance to Biodiversity will be discussed as an example in this chapter.

Keywords: *Biodiversity, Genetic resources, Traditional knowledge, Intellectual property, Patent, inventories, Catalogues, Panama.*

Introduction

A major concern for developing countries is the fact that the intellectual property (IP) rights system tends to increase the costs of access to many products and technologies in areas of particular developmental importance like health, agriculture, education and information technologies (Anonymous, 2002b).

Like many other developing countries, Panama has reacted protecting its genetic resources and traditional knowledge. However, in spite of the efforts of the government to demand that foreign research disclose the geographic source of the genetic material from which the "invention" is derived, the country is unaware of many of the studies its resources has been evaluated for.

In this matter, Panama has followed the lead of countries like India, China and Brazil, in cataloging its traditional knowledge and preparing inventories of its genetic resources, in order to check claims of invention against existing traditional knowledge.

As signatory of the Convention of the Biological Diversity (CBD), Panama encourages the objectives of the Convention: conservation of biological diversity, sustainable use of its components and the fair and equitable sharing of the benefits arising out of the utilization of genetic resources (Anonymous, 1993) and it is aware that intellectual property rights "may have an influence on the implementation" of the CBD and that it is compulsory for states to cooperate in order to ensure that intellectual property rights are "supportive of and do not run counter to" the objectives of the CBD (Anonymous, 2000a).

Nevertheless, it is evident that the interaction between the rights referred to in TRIPS (World Trade Organization's Agreement on Trade-Related Aspects of

Intellectual Property Rights) is contemplated as well within the CBD. Concerns such as software, which relates to conservation and information that relates to sustainable use of biodiversity may be the subject of copyright protection (Anonymous, 2002b). In our particular case, much scientific information related to conservation and sustainable development is found in databases of restricted access. The government is pushing, through the National Secretariat of Science and Technology (SENACYT) to a more open access to this information, especially to governmental institutions focused on regulation and protection of the national biodiversity, as well as those of academic scope.

The present work presents the view on Patent system in Panama, mainly focused on the subject of protecting biodiversity and traditional knowledge as it is an important national heritage.

The System in Panama

The Legislation

Law 35 of May 10th, 1996 is "designed to protect inventions, the models of use, the industrial models and drawings, industrial and commercial secrets, trademarks for products and services, collective and warranty trademarks, the trademarks of origin, origin denominations, commercial names and advertising slogans and signs" (Anonymous, 1996).

Later, the Legislative Assembly published Law 20 of June 26th, 2000 destined "to protect the collective rights of intellectual property of indigenous communities upon their creations such as inventions, models, drawings and designs, novelties contained in the pictures, figures, symbols, illustrations, old carved stones and others; likewise the cultural elements of their history, music, art and traditional artistic expressions, capable of commercial use, through a special registration system, promotion, commercialization of their rights in order to stand out the values of the indigenous cultures and to apply social justice" (Anonymous, 2000b).

On April the 29th, 2009, the President of the Republic of Panama signed the Executive Decree No. 25 with the purpose of "regulating the access to genetic and biological resources in which Panama is the country of origin, from their derived products, and the genetic resources of the migratory species that by natural causes are found in the territory of Panama; as well as the procedures to access knowledge, innovations and traditional practices associated to the use of genetic and/or biological resources, and the benefits derived by commercial and other type of uses, of those resources, excluding human genetic resources" (Anonymous, 2009). Through this decree, the National Authority of the Environment is capable of regulating access and intellectual property of genetic resources.

The major drawback of these legislations is that even though they cover the most important aspects like: patents, obligatory licensing, certificates of origin, letters of compromise, prior informed consent; the section intended to cover the regulation of traditional knowledge originated by the use of genetic resources is missing, so that traditional medicine have been left out.

Documenting Resources and Knowledge

The direct impact of IP protection is to benefit financially those who have knowledge and inventive power, and to increase the costs of access to those without. If most developing countries do not have a strong technological base which could benefit from IP protection, they do have genetic resources and traditional knowledge, which have value both to them and to the world at large (Anonymous, 2002a).

Panama has reacted to the emerging IP system by creating units within governmental institutions in order to protect Panamanian genetic resources and traditional knowledge. Following the lead of countries like India, China and Brazil, Panamanian scientists are collaborating with the authorities in cataloging its traditional knowledge and preparing inventories of its genetic resources, in order to facilitate to check claims of invention against existing traditional knowledge. However, in spite of the efforts of the government to demand that every research disclose the geographic source of the genetic materials from which the "invention" is derived, in many cases this information is incomplete.

Softwares

In Panama, scientific information related to conservation and sustainable development is found in most cases in databases of restricted access. Moreover, several research and academic institutions in Panama have data bases on Panamanian biodiversity with different formats. The Institute for Scientific Research and High Technology Services (INDICASAT-AIP), with support from the National Secretariat of Science and Technology (SENACYT) in Panama, made this year the first attempt to standardize the existing data bases in the country into one format, taking as a model Biotica®, a data base created and used by the National Commission for the Knowledge and Use of Biodiversity (CONABIO) in Mexico to manage data related to taxonomy, geographical distribution and ecological parameters of biodiversity (http://www.conabio.gob.mx/biotica/cms/index.php). This initiative is expected to initiate collaborations that will allow us to have a more open and easy access to this information. Particularly to those institutions focused on the regulation and protection of the national biodiversity, as well as those of academic scope.

Catalogues

The cases like invalid patents issued because the patent examiners ignored the relevant traditional knowledge or cases were countries were entitled to receive a fair share of any benefits arising from commercialization of a plant identified within their traditional knowledge, are the main reasons why the international community is aware of the need of documenting traditional knowledge (Anonymous, 2002a)

Indigenous knowledge and the collective rights of the Amerindian tribes in Panama has been covered by Panamanian law No. 20 of June 26th, 2000 of the Legislative Assembly, where the customs, traditions, folkloric expressions, traditional knowledge and other type of traditional expressions, cannot be object of any form of exclusive right by not authorized third parties under the intellectual property system (Anonymous, 2000b). However, this law has a deficiency, it cannot be applied to traditional knowledge associated to genetic resources, it is only applicable to knowledge related to handcraft, dances, music and customs, that is why the Executive Decree No.25 is adopted in the context of prior informed consent and the contract of benefit sharing. This way there is still a legal tool to protect them from the biopiracy and to guaranty economic benefits derived from the use of this knowledge. The decree has yet one flaw, this law is applicable only to knowledge from indigenous tribes excluding local peasants and afrocaribbean communities.

Panama possesses five Amerindian groups: Ngöbe-Buglé, Kuna Yala, Emberá-Wounan, Naso or Teribe and Bri-Bri, whose traditional medicine knowledge is very rich, but it remains in its majority as unwritten knowledge (Figure 12.1). A fraction of this information has already been published by a research group coordinated by Dr. Mahabir Gupta at the University of Panama (Gupta *et al.*, 2005; Gupta *et al.*, 1993; Joly *et al.*, 1990; Joly *et al.*, 1987) an it is available on line.

Panama has also participated in the regional initiative called TRAMIL-Centroamerica (Program for Scientific Applied Research and Popular Use of Medicinal Plants in the Caribbean, coordinated in Panama by Dr. Pablo Solís), whose objective is to contribute to biodiversity conservation using the popular knowledge and scientific evaluation of medicinal plants, in close collaboration with ONGs, state and academic institutions. This organization has published the results of this work in Journal Vegetal Caribbean Pharmacopeia (TRAMIL, 2005).

Figure 12.1: (Left) Baby from the Tribe Emberá is Covered on an Aqueous Solution of the Fruits of *Genipa americana* L. to Treat Skin Problems; (Right) Woman from the Tribe Gnöbe-Bugglé Teaches the Community How to Use *Aloe barbadensis* (Miller) to Treat Burned Skin.

Nevertheless, the information that has been made available is summarized due to the limitations of journal publications. So the need of a complete accessible written record remains.

Inventories

Numerous inventories have been done on Panamanian biological diversity. Inventories on plants, mammals, birds, reptiles and many other organisms have been well documented in books (Ibanez *et al.*, 2005), scientific journals (Collin *et al.*, 2005) and kept in data bases (http://biogeodb.stri.si.edu/bocas_database/). The problem arises when, in spite of the efforts of the government to demand that research discloses the geographic source of the genetic material from which the "invention" is derived, in many cases geographical indications are obviated and the country becomes unaware of the whole study its resources has been evaluated for.

Other Considerations

Both IP rights and CBD recognize that biodiversity-related matters deal primarily with patents. In biotechnology in Panama, product patents have been granted on, DNA sequences, genes, the products of genes and micro-organisms as long as there is an application or use described. Animals and plants are not subject for patent according to national legislation; and transgenic plants can be protected by the International Union for the Protection of Vegetal Varieties (UPOV for its Spanish acronym).

Developing countries, however, have been discourages from providing patent protection for living organisms and its products, as is allowed under TRIPS, because of the restrictions patents may place on use of these organisms by farmers and researchers (Anonymous, 2002b).

Technology Transfer is a major concern and is not yet well established in Panama. There is growing encouragement to transfer indigenous technologies to be developed at INDICASAT-AIP. We are mainly looking for low cost technologies.

Research and Development: The research related to drug discovery, indigenous herbal medicines, low cost health kits, vaccines, diagnostics are popular research areas in Panama. Many times, important discoveries are made public through press meet even before going for patenting or publications. Now we are trying to create awareness on IP rights and its governing regulations.

Plants, Intellectual Property Protection and farmer´s right: It is still in infancy stage in Panama. This is mainly concerned with reference to herbal medicinal knowledge base, local unique crop based and traditional cross breeding to develop new commercial plants.

Conclusions

In this article we identify several national priorities that will contribute to protect and promote traditional knowledge. The first is by cataloguing traditional medicine from our Amerindian groups, which we propose to do gradually starting from the Ngöbe-Buglé community which we have made the first contact with, and following the standards presented by India in the Traditional Knowledge Digital Library. The challenge will be to have all the tribes to agree to be incorporated to this system.

Controversy remains, whether there is a need for an Intellectual Property System dedicated to protect traditional knowledge coming from the use of genetic resources (like indigenous traditional medicine) or if the indigenous populations will want to be part of it.

Scientific information related to conservation and sustainable development should always be found in databases of open access, so that governmental institutions can use them to better design laws to protect national biodiversity. We propose the use of a uniformed data base format, which can incorporate the information of existing individual data bases, facilitating and speeding up the access to this information.

We believe that only by properly documenting our own biological diversity and traditional knowledge, we will be able to efficiently protect it.

Panama concur that developing countries need a combination of increased international funding, specific initiatives to support research into areas that lack a lucrative market (neglected diseases like malaria, dengue, etc), and support for increased education, technology transfer and training within developing countries.

Acknowledgements

The authors thank Dr. Rubén Berrocal, National Secretary for the National Secretariat of Science, Technology and Innovation (SENACYT) for all his encouragement. The authors are also grateful to Dr. Mahabir Gupta for contributing with pictures and data on Panamanian folk medicine and for all his support during course of investigation.

References

Anonymous, 1993. United Nations Conference of Environment and Development. Convention on Biological Diversity. December.

Anonymous, 1996. Legislative Assembly. Law No. 35 of May 10th 1996. Whereby provisions on industrial property are adopted. Republic of Panama.

Anonymous, 2000a. CEAS Consultants (Wye) Ltd, Centre for European Agricultural Studies. Study on the Relationship Between the Agreement on Trips and Biodiversity Related Issues, September.

Anonymous, 2000b. Legislative Assembly. Law No. 20. of June 26th 2000. On the special intellectual property regime upon collective rights of indigenous communities, for the protection of their cultural identities and traditional knowledge, and whereby set forth other provisions. Republic of Panama.

Anonymous, 2002a. Commission on Intellectual Property Rights. A call to improve intellectual property rights of developing countries. October.

Anonymous, 2002b. Commission on Intellectual Property Rights. Integrating Intellectual Property Rights and Development Policy. London, September.

Anonymous, 2009. Ministry of Commerce and Industry. Executive Decree No. 25 of April 29th 2009. WHEREBY ARTICLE 71 FROM LAW NO. 41 OF JUNE 1, 1998, General Law of the Environment, IS REGULATED. Republic of Panama.

Collin R., *et al.*, 2005. Photographic identification guide to some common marine invertebrates of Bocas del Toro, Panama. Caribbean Journal of Science, 41(3), pp. 638-707.

Gupta, MP. *et al.*, 2005. Medical ethnobotany of the Teribes of Bocas del Toro, Panama. J Ethnopharmacol, 96(3), pp. 389-401

Gupta, MP. *et al.*, 1993. Medicinal plant inventory of Kuna Indians. Part 1. J Ethnopharmacol, 40(2), pp. 77-109

Ibanez A. *et al.*, 2005. Catalogue of the flora of the Bahía Honda region (Veraguas, Panama). In: Santiago Castroviejo;Alicia Ibanez (Ed.), Studies on the biodiversity of the Bahia Honda region (Veraguas, Panama): 177-317. Madrid: Consejo Superior de Investigaciones Científicas.

Joly, LG.*et al.*, 1990. Ethnobotanical inventory of medicinal plants used by the Guaymi Indians in western Panama. Part II. J Ethnopharmacol 28(2): 191-206

Joly, LG.*et al.*, 1987. Ethnobotanical inventory of medicinal plants used by the Guaymi Indians in western Panama. Part I. J Ethnopharmacol 20(2): 145-171

TRAMIL, 2005. Farmacopea Vegetal Caribena. Editor L. Germosén Robineau, 2ª. Ed. Act. Editorial Universal, UNAN-León, Dominican Republic, pp. 486

http://www.conabio.gob.mx/biotica/cms/index.php

http://biogeodb.stri.si.edu/bocas_database/

Chapter 13

Research and Development Networks in Colombia: An Answer for the Research Ownership Exploitation Challenge

Maria Consuelo Velasquez

I.P. Advisor,

Columbian Institute for the Development of Science and Technology (COLCIENCIAS),

Bogota, Colombia

E-mail: mcvelasquez@colciencias.gov.co

ABSTRACT

Based upon the facts that were taken into account by the Applied Research Project that took place from 2004 to 2006 supported by the World Intellectual Property Organization (WIPO), the national authorities realized that Colombia, as well as most of the developing countries, had excellent researchers, but lacked infrastructure and financing. Besides, the researchers were not aware of how to protect the results of their investigations and even more, how to profit from them. Even though the above mentioned study was focused on the health sector, the Administrative Department of Science and Technology of Colombia (COLCIENCIAS) considered that other economic sectors presented the same situation, such as the Research Ownership Exploitation Challenged. In Colombia the researchers generally do not own the results of their work and, what is even worse; they are not even entitled to commercialize it, therefore are unable to realise economic returns on Research and Development investment. The situation discourages researchers doing research and investments in Research and Development. Looking for an answer for the Research Ownership Exploitation

Challenged and following the objectives of the Applied Research Project, and using the scheme created in the health sector, since 2008 in a jointly effort with the Superintendence of Industry and Commerce (Intellectual Property Office), COLCIENCIAS had been working in a project named: Research and Development Network in Colombia. Consequently, creating four Networks compromised in achieving goals of the Government of Colombia, such as to create the conditions for the knowledge to become a tool of the development.

Keywords: *Intellectual Property Networks, Networks in Colombia, Research, Research and Development, Research results ownership, Research results ownership exploitation, Hub.*

Introduction

The main objective of this document is to present what in the author's opinion, could be a solution to the "challenges of ownership entitlement and exploitation of research results", that are present in the research field in Colombia. It draws on the author's personal experiences as Intellectual Property Advisor in the Administrative Department of Science and Technology[1], for now and on COLCIENCIAS, and it reflects the author's personal conclusions.

The document consists of four parts. The first one is a brief summary of the activities that have been done in COLCIENCIAS, regarding intellectual property. The second part, describes a project that constitutes the pilot base of the scheme in which we are currently working. In the third part, the first two stages of the current project are discussed. The paper concludes with the fourth part with a proposal for further continuity.

Background

Knowledge is a means to lead the country on the path of Economic Development. The government of Colombia, in its Vision for 2019, on Science, Technology and Innovation, aims to "produce, disseminate, use and integrate knowledge to contribute to productive and social transformation of the country. For this, it has created the National Development Policy for the Promotion of Research and Innovation, whose main objective is "creating the conditions so that knowledge would be a development tool".

COLCIENCIAS is one of the entities called upon to achieve this goal, within its Administrative Department of Science and Technology role; and as head of the National System of Science and Technology, has presented a series of strategies that focus on the good use of national resources and talent. Strategies that seek to expand economic opportunities in a globalized world that encourages the participation of the different stakeholders (academic, business and government), in the creation of policies and programs for better cooperation between the three; to make real progress in building more favourable conditions for the development of the country. For the

1 The entity in Colombia that is in charge of guiding and promoting the National Scientific and Technological Development.

aforementioned, and in search of accomplishing its mission objective COLCIENCIAS has done, activities regarding to the subject of Intellectual Property; which at last stimulate creativity to improve the quality of life for the country.

In 1997 under the technical direction of the National University of Colombia, published the: "Handbook on Intellectual Property products derived from the academic activities in universities and research centres". It counted with the participation of experts in each of the branches of intellectual property.

In the year 2005, with the aim of filling the lack of financial aid given to the creators; with the support of the Superintendence of Industry and Commerce, the first call to fund patent applications, in Colombia and abroad, was managed. All this, within COLCIENCIAS clear mission to facilitate the ownership of knowledge generated by legal representatives and individuals dedicated to research.

In the year 2006, to raise awareness among entrepreneurs, researchers and interested public, on the basics of intellectual property, in collaboration with the Superintendence of Industry and Commerce, the National Directorate of Copyright and the Colombian Institute of Agriculture, ICA, organised eight workshops on these topics. In 2007, the training continued with in depth emphasis on patents, with the support of the Superintendence of Industry and Commerce.

Simultaneously, and to complement this type of training in 2007 COLCIENCIAS jointly with a private foundation, held workshops for the negotiation and valuation of technology, that sought to train technology negotiators. Those negotiators would serve the purposes of the National System of Science and Technology and its stakeholders: companies, research and technology development, universities and other organizations, for the incorporation and transfer of new knowledge for developing successful businesses and their ability to compete.

In the year 2008 the National Council for Science and Technology established through the 8th Agreement, that the Intellectual Property arising from the contracts and agreements accomplished by COLCIENCIAS and other entities of the National Science and Technology System would now belong to the executing agencies.

Nevertheless, the mayor project related to Intellectual Property in which COLCIENCIAS has yet participated is the "Networks of Intellectual Property Management for Strengthening of Research Capacity", is still ongoing and has been able to meet the main objectives. Regarding this project, the author presents what has been achieved in the first two stages, including a proposal for the implementation of a third stage.

Pilot Project "Network for Research and Intellectual Property in the Area of Health"-WIPO (World Intellectual Property Organization)2

The project called "Research and Intellectual Property Networks Project: A model for Supporting Developing Country Researchers in the Creation, Protection and

2 For now on WIPO.

Exploitation of Research Results in the Field of Health ",was coordinated by the World Intellectual Property Organization, with the participation of COLCIENCIAS and the Superintendence of Industry and Commerce[3] as local counterparts.

Main Problem

The WIPO within its vast experience has identified a problem in developing countries that plagues common institutions engaged in research. Despite the quantity and quality of researchers extremely dedicated to research, lack of appropriate infrastructure and funding, and find it difficult to obtain the ownership and exploit the results of their research using the Intellectual Property System, because of the scarcity of resources, infrastructure, training and legal and professional marketing services.

Due to these difficulties, researchers do not hold the ownership over the result from their research, and cannot commercially exploit it because they do not know how to do it. Frequently, investment in research and development has no economic returns, and social benefits resulting from research and development activities are limited.

These events often discourage the authorities of developing countries from investing in research to solve problems they face.

This situation, which we can name "the challenge of ownership and research result exploitation", caused by a series of interrelated factors that provoke a vicious cycle:

- ✰ Research and development institutions of developing countries are often not fully aware and lack the resources necessary to protect the results of their research, acquire ownership of them, and exploit them as intellectual property derived from said research results. There is also a serious shortage of patent drafters. Even more, there are very few people who can negotiate and draft licensing agreements, and provide marketing facilities and technology in relation to intellectual property.
- ✰ Research and development institutions often lack the policies and procedures governing intellectual property at institutional levels, such as, mixed public –private partnerships, sponsored research, disseminating research results and economic incentives for researchers.
- ✰ The low level of Investment in research and development or even cuts in this Investment, can threaten levels of research results and consequently, reduce the chances of retaining and attracting the most talented researchers. As in a spiral, lack of financial and social returns derived from research and development may in turn lead to further reductions in financial support by the public sector institutions of research and development.
- ✰ When negotiating technology transfer agreements, the lack of acquisition of ownership of research result can contribute to a disadvantage in research institutions of developing countries, because they do not possess exchange tradable assets.

---

3 Intellectual Property Office in Colombia.

Objectives

The main objective of the project was to support Researchers in the field of health in French-speaking Central Africa and of Colombia, to protect and exploit the results of their research and thus, promote intellectual property of the results of their research and exploit them commercially, if they so decided. It is also proposed to build a network of research centres and universities in intellectual property taking advantage of skilled human resources. The expected results of this September 2006 project were the following:

- ☆ To put into operation two networks of research and development in the field of health, with the support of two separate Intellectual Property Shared Serviced (I.P.S.S); the researchers will be more capable of make decisions regarding the protection and commercial exploitation of research results.
- ☆ To have a group of people from developing countries, who received training in three key areas of Intellectual Property and will be an acquisition to offer their knowledge to their respective scientific communities.
- ☆ To form diverse interdisciplinary teams working together to seek solutions to common challenges of Researchers from developing countries ("the challenge of ownership and exploitation of research results"), forging links and sharing experiences.
- ☆ To develop, implement and evaluate a model for strengthening the research and development performed in developing countries; a model that will be available to be shared internationally with the scientific community and with the government officials responsible for science, technology, health, and intellectual property.

Development of the Project

The steps followed to achieve the objectives mentioned as above were the following:

- ☆ Audit;
- ☆ Development and delivery of the training program consisting of three modules;
- ☆ Implementation of Intellectual Property Shared Services.
- ☆ Survey and evaluation of the model developed.

Stage1: Audit
Analysis of the Situation of IP in the Participating Institutions

At this stage, an audit conducted in research and development institutions. Interdisciplinary teams must study the situation in which these institutions are, regarding planning and management of health research in those areas that might affect the protection, ownership and exploitation of research results. They will consider, among other things, the following elements: sources of financing available for institutions to pursue their health research; institutional policies and procedures regarding the protection and exploitation of the results of research and development through patents and other forms of intellectual property. They have to study the results of currents research that could require immediate protection; availability of personnel capable of marketing and commercialization applied to the results of protected research. Besides, is important to create measurement indicators to apply them throughout the project.

Colombia

Within the project in Colombia, the audit stage focused on the determination of a diagnosis on the treatment given by the institutions that are part of the project of protection, management and commercialization of research results in the area of health. Therefore, the team will know the reality of such institutions and focus the training of the second stage according to the maturity or the management of the intellectual property issues.

The audit based on a special questionnaire that reflected all the issues necessary to study intellectual property within an institution that performs research and development. Before understanding the general reality of research institutions, the first meeting of the project brought to light that "it could be established by the assistance that it was the first time that the worlds of research and development and intellectual property united and that the project was worthwhile".

The conclusions reached after studying the responses to the questionnaire of the audit were as under:

- ✰ The R&D institutions lack of awareness and knowledge on the subject of intellectual property.
- ✰ The level of definition and implementation for institutional policies and procedures of intellectual property is very low.
- ✰ Lack of trained persons in the areas of patent drafting, licensing agreements, and management and commercialization of research results;
- ✰ Lack of appropriate institutional infrastructure, to protect and commercialize research results;
- ✰ Tendency to publish without considering the protection by other means;
- ✰ Minimal protection of research results;
- ✰ Minimal allocation of resources for protection, management, and commercialization of research results;
- ✰ Absence of negotiations to license technology and knowledge;
- ✰ Little economic return on the investments made in research and development.

Stage2
Training Programs

Considering the results obtained by the audit, an interdisciplinary team must prepare a curriculum composed of three modules of training in key practice areas for the management of intellectual property assets in the field of health research, namely:

- ☆ Management of health research in the light of intellectual property and commercialization of its results;
- ☆ Training of Patent drafters; and
- ☆ Training on license agreements.

Once developed, each of the three modules of the training program will be offered to a selective group of fifteen people in each of the two geographical regions of the project (for example: 15 people x 3 modules = 45 people in the French-speaking sub-region of Central Africa, and the same number in Colombia.) The criteria to select people who will attend the training program was the prior knowledge relevant to the best employ of each module, the intention to apply the acquired knowledge and the real possibilities of implementing them given their current work situation. The majority of the beneficiaries of the training will come from research institutions participating in the project.

Training Program in Colombia

Colombian institutions understood from the outset of the formulation of the project that training was a key element in achieving the objectives. Therefore, to develop this activity, we designed three training modules, including the organization and distribution of technical material and accompanying disclosure experts in the respective subjects, that was not easy for the issue of management and marketing health research results.

- ☆ Training of Patent drafters;
- ☆ Licensing agreements module; and
- ☆ Management and commercialization of health research results module.

Groups of people who are currently in the research and development institutions of the health sector trained throughout the country putting into practice the knowledge acquired and notwithstanding the above. Because the project focalized in the health sector, it was necessary to extend the project to other sectors in order to find the agreement at institutional levels, because often the work of these people vanished due to the lack of uniformity in the policies of intellectual property management.

However, many of the directors of research institutions attending the training allowed the discussion of important topics within the project "Research Networks and Intellectual Property" such as:

- ☆ The contents of the agreement that was signed between the twelve institutions of the project for the formal establishment of the Intellectual Properties Shared Services;
- ☆ The importance and role of institutional intellectual property policies at universities and research centres to try to establish, for example, ownership, incentives and procedures, particularly for the protection, management, and commercialization of research results.

Stage 3
Launching the Two Networks of Research and Development and their Respective Shared Services of Intellectual Property

The I&D Institutions and universities involved into the project will receive information and technical assistance on institutional policies, procedures, forms and contracts related to the governance of research results and research and development assets therefore. They will also receive information on models of institutional policies and agreements for the creation of research and development based networks and intellectual property advice on the opportunities and options for creation.

For the establishment of the two IP Shared Services, one for the Network of Central African Intellectual Property and one for Colombia, three participants who have obtained the highest grades in the modules training will be part of the hub in each of these geographical areas. All the participants in the training program will continue to receive advice and support from experts and staff responsible for the development and implantation of the three modules as well as the support of OMPI. The three participants selected in each of the two areas will be recruited by the project to form the corresponding IP Shared Services that institutions will provide to their respective network related to intellectual property. These three members of each IP Shared Services also will provide support in the process of creation and consolidation of the respective network between institutions of research and development, and will study and monitor the model by the project to provide information for the preparation of the assessment study. Other participants will return to work in their respective institutions and put into practice the acquired knowledge.

Launching the Network in Colombia

In Colombia, the implementation of the health hub became operational as a scheme of interagency cooperation to create an outline of IP Shared Services as part of the network. Because that scheme provided the institutions of the network with services for protection, management, and exploitation of research results, and proved that, their operation is essential for the support and for the strengthening of innovation. In this sense, knowledge achieved, becomes the basis for implementing the project in other participating institutions and creates new institutional capacities.

Stage 4
Evaluation Study of the Model Developed

An interdisciplinary team will conduct the study model developed under the project, the member must come from the project partner institutions in Switzerland, Central Africa, and Colombia, and in addition with experts and teachers that include a researcher (preferably a graduate or post-graduate student) selected in each of the three geographical areas. Researchers from Central Africa and Colombia will act locally as "observers of the project" and investigate, monitor and give the necessary assistance to the project on the ground. The multidisciplinary team responsible for the evaluation study will also receive the support and input from the research, monitoring and evaluation undertaken in the field component of the IP Shared Services for Central America and Colombia.

At this stage, activities of analysis, evaluation and measurement based on indicators developed during Stage 1 of the project will be included to estimate the impact of the model in aspects such as applied scaled economies for the networks, possible repercussions of the model on research acceleration and greater familiarity with the intellectual property system. They will measure the inclusion of data on the use of forms of disclosure of inventions and other aspects related to the management of research and development results, questionnaires and reports of satisfaction of the users of IP Shared Services, difficulties identified in the implementation of the network and IP Shared Services, etc. The evaluation study will also include strategy for sustainability-oriented recommendations to make a link between research and development and IP Shared Services and to continue to function after September 2006, date of completion of the project.

Evaluation of the Model in Colombia

The project is still in development stages.

Special Project "Networks of Intellectual Property Management for Strengthening the Capacity of Research"4

General Considerations

Definition

Based on the pattern learned in the pilot project funded by WIPO in the health sector, COLCIENCIAS accompanied by the Superintendence of Industry and Commerce has been developing the project: "NETWORKS FOR RESEARCH AND INTELLECTUAL PROPERTY MANAGEMENT". The long-term view of the project is to achieve the creation of a real culture of intellectual property and its mission has been to boost the construction of virtual knowledge communities, or communities of practice. These communities could raise the scientific and technological capacities of the different actors of research and innovation, and can streamline investments in human resources, infrastructure and take advantage of the cooperation achieved.

In the context of the Science, Technology and Innovation Policy, in the line of support for intellectual property, developed in the document regarding to Intellectual Property (CONPES), it seeks the creation of networks of Research and Development Centres in areas of science and relevant technology, in production lines or regions.

Networks will have, as basic pillars, the Research Centres, Universities, and Technology Development Centres, conceived as organizations, specializing in research generation and transfer of knowledge and technological services. The primary mission of the networks will be to boost the knowledge market in Colombia through IP networks under a scenario of a more productive and globalized economy.

The purpose of networking is to move forward in building knowledge communities, which can raise the scientific and technological capacities of the different stakeholders of research and Innovation, and rationalize Investment in infrastructure and human resources.

---

4 Project that the author is currently coordinating.

Networks arise from the interest of its members, by the affinity of the technologies and knowledge for development, or by the demand conditions for business services organized through cooperation and interaction patterns that defined in terms of sustainability and growth of their members.

Networks organized through agreements, temporary unions, or other associations among its members, with formulas that do not involve legal or organizational changes within each of the centre or institution that composed them. The definition of the form of organization and integration of the notes and of the leading centres arise from the actions of the networks.

Research and development networks are the result of collaboration among research institutions in which each institution agrees to take and implement certain policies common to all members of the network and even share some services. By applying models of scaled economies and availability of certain resources and shared services, networks can help to reduce costs and may even encourage the acceleration of research. COLCIENCIAS, aware of the importance of its role as the Administrative Department of Science and Technology, has focused on supporting the creation and implementation of a real culture of intellectual property, stipulating the following:

Objectives

The project aims to support Researchers in the protection and exploitation of research results. To achieve this objective, the project will research, develop, implement and evaluate a transferable model for managing intellectual property assets. This model based on the development of local experts on the management of intellectual property, research and development networks and IP Shared Services that will help network members to take advantage of their research and development results and contribute to the sustainability of these efforts.

General

- ✰ To advise the participating institutions in the protection, management and exploitation of the results of their research and development through training and technical assistance.
- ✰ To form a model network of research and development through IP Shared Services between members.
- ✰ To develop protocols for technology transfer offices in selected institutions

Specific

- ✰ To promote alliances between public and private sectors
- ✰ To promote the relationship with the productive sectors
- ✰ To utilize strategies for the commercialization of intellectual property
- ✰ To fundraise and research funding

Stages

Stage 1: Audit

Stage 2: Development and Training

Stage 3: Implementation of the Networks

Stage 4: Survey and Evaluation of the Model

Stage 1: Audit: Analysis of the Status of Intellectual Property in the Participating Institutions

This activity focused on the determination of a diagnosis regarding the issues of protection, management and commercialization of research results in all the selected sectors, in order to get a closer look to the reality.

For the audit, they create a special questionnaire to reflect all of the relevant issues related to the intellectual property within an institution that performs research and development. Before getting to know the general reality found in research and development institutions, the attendees of the audit at the first meeting of the project brought to light that it was the first time the worlds of research and development and intellectual property were joined and that the project effort was worthwhile.

This audit process produced the following results in the different sectors:

Agro-Industrial Sector

- 12 entities evaluated;
- 4 identified projects;
- 12 members.

Health Sector

- 20 evaluated entities;
- 2 identified projects;
- 12 members with institutional policy.

Industry and Energy Sector

- Diagnostic of research groups with the potential of IP;
- Between 10-25 projects for application were found;
- 10 institutions signed the agreement.

Defence Sector

- 20 evaluated entities;
- Results pending.

Stage 2: Training

According to the Document CONPES[5] 3533; the role of COLCIENCIAS is to support the design and implementation of a national program aimed to promote industrial property as a mechanism for business development. For this, and following the recommendations of the same document (a), it is essential to continue making ongoing training activities through workshops, meetings, seminars, forums, and other courses to achieve the knowledge of industrial property and contribute to the

5 Document that contains the basis of the public policy decisions to be taken trough each the presidential period.

formation of a critical mass that will use the issue as a tool for obtaining base development knowledge for the country.

Therefore, the training stage of the project consists on ongoing process that contributes not only to create awareness in the community, but also to give support to the creation of a real culture of intellectual property in the country. More specifically to give continuity to the second stage of the project that will be an entry for new members (including the private industry). Besides that establish a system of continued education that should include at least the following modules:

- Introductory Model
- Technology Intelligence
- Commercialization
- Patent Drafting

Stage 3: Implementation of the networks:

Strategy 1 of CONPES 3533, follows the recommendations (b), (c), (d); in which strategies are recommended to promote the adoption or adequacy of internal regulations on intellectual property in research centres that include aspects of ownership and distribution of economic benefits. This is perfectly consistent with stage 3 of the project "NETWORKS FOR RESEARCH AND PROJECT MANAGEMENT OF INTELLECTUAL PROPERTY" that has been launched to consolidate the network as well as the respective IP Shared Services (IPSS). In those research and development institutions and universities, the support of institutional policies, procedures, forms and contracts related to the governance of research results and intellectual property assets derived there from are necessary.

The four networks are ready to launch the third stage of the project called Strengthening Networks, and its main objective is to achieve an institutional support for Institutional Policy and Intellectual Property and advice on legal procedures to protect research results.

Results (Updated June 2011)

Agro-Industrial Sector

- 3 Patent applications under review
- 2 Apellations of Origin Granted
- 2 Apellations of Origin applications
- 6 Breeder´s Rigth applications
- 2 Utility Model applications

Health Sector

- 11 Patent applications

  9 pending before the Intellectual Office in Colombia

  2 pending before PCT

Industry and Energy Sector

- ✰ 15 Patent applications
  - 9 pending before the PCT
  - 2 under lawyer review
  - 4 in progress

Panorama

There are four networks in Colombia created by universities and research centres that are part of the hub. The unifying criterion that qualifies these networks is the sector they belong to, namely Agriculture, Industry and Energy, and Defence. Within the project, some universities develop the protocols for their technology transfer offices and settle the basis of IP statues.

During the development of the second stage, we found that some institutions were not yet able to identify the commercial value of the results of their investigations and lack the resources to protect, acquire ownership, and exploit these results.

Some of the causes that have been identified are:

- ✰ No trained patent drafters, no trained negotiators and dispersed knowledge over drafting license agreements, and marketing services in relation to intellectual property.
- ✰ Few resources devoted to R&D activities.
- ✰ No continuity in the training program.

Taking into account the objectives originally proposed, the conclusion is that the hubs of Agro-industrial, Industry and Energy, and Health Sectors have completed the first stage and the performance of the second stage has been relatively successful. Nevertheless, the Defence Sector Network has advanced the first and second stages simultaneously. Because of this, it is the time to launch the third stage of the project where the institutions will receive guidance on policies for managing intellectual property. However, it is important to make some adjustments that involve continuing with the second stage, within a training program targeted specifically to interdisciplinary groups that are responsible for providing shared services.

Proposal

Extension of the Second Stage

It is required that each one of the institutions, that are part of the hub, designate an interdisciplinary group composed of at least two professionals. This people must attend the three modules of the training program, will implement stage three in their institutions and are responsible for providing adversary services to members of the network, which will become apparent compliance with the objective of IPSS for the network to materialize.

The purpose is to continue with the training making stage two into a continuous stage. OMPI cooperation will be essential in the achieving of this goal. The continuous cycle will have at least the following topics, according to the expressed needs of the coordinators of the hubs in each of the four sectors:

- Basic Module
- Valuation of Intangibles Assets
- Negotiation of Technology
- Patent Drafting

Third Stage (Year 2011)

Research and development institutions and universities involved in the project will receive information and technical assistance on institutional policies, procedures, forms, and contracts related to the governance of research results and IP assets. There from they will receive, as well, information on models of institutional policies and agreements for the creation of research and development based networks and intellectual property advice on the opportunities and options for creation.

Consolidation of IPSS

IP Shared Services provide institutions outside the research development network with protection services, management, and exploitation of research results and consequently its performance is vital to support and strengthen Innovation. These IP Shared Services include:

- The support of the formalities related to the legal protection of results of research and development and therefore it facilitates the appropriation of its ownership through the use of the IP system;
- The management of IP owned by institutions of research and negotiation, and drafting of licensing agreement based on this IP;
- The promotion of joint public-private agreements;
- The use of applied online marketing and the commercialization of their intellectual property assets;
- The acquisition and negotiation of funds and financing;
- The promotion of the development of local manufacturing.

The hubs have in their plans, to continue training activities with the joint participation of the four networks for the fulfilment of the specific demands of each of them. They want to achieve the implementation of a tool for communication and dissemination, with the participation of the four sectors, which will contain a public part that will be able to put information from each one of the sectors. That virtual tool will be useful to meet the assessment needs for the management of intellectual property.

Policies

One issue that has been dormant for a long time and has gained force in recent years in the country is the need for the existence of policies of intellectual property management in universities. It is suitable that these policies are supported by the state and that they can be applied in public and private universities so that they become tools of Technology development for the country by successfully linking the academic sector with the productive sector. Likewise, it is required that development of incentives that lead to the implementation of these tools in the activities of universities and research centres.

For the aforementioned, it is necessary to include universities that belong to networks in the project of integration along with the Ministry of National Education, and other public institutions, to achieve uniformity in policies of Management of Intellectual Property. This work also considers the same criteria to achieve a similar policies for R&D institutions.

Conclusion

Colombia has great potential in research and in each of the sectors it has good prospect to work on intellectual property issues, for example, in the area of health, research and development networks. It can play an important role in promoting the production and local distribution of medications based on research developed using both conventional approaches as well as traditional medicine. In addition, seeking partnerships supported by the government to insure that research aim to the needs of the pharmaceutical industry in countries like ours (tropical diseases). In the agricultural sector, the infinite possibilities of our bio-diversity force us to take a position on the access to genetic resources and there are infinite possibilities for plant breeders. The industry and energy sectors are where theory that Intellectual Property should be a tool for development truly materializes. In addition, in the Defence sector, creativity originated in the situation of our country and the need to regulate the management of resources arose from the Agreements of Compensation- Offsets- that creates the unusual vigour of intellectual property.

Yet the only way to ensure that research becomes a tool that leads to the development of the country and meets the challenge of the ownership and exploitation of research results, is to give researchers the ability to adequately protect results of their research and make alliances with the productive sector so that the system becomes innovative. This will bring more motivation to continue researching. The state is responsible for building the bond that unites the academic sector with the business sector.

References

National Department of Science and Technology (Colciencias): Colombia plants and builds future. National Development Policy of research and innovation. American forms and Print. August 2008. ISBN 789588290461.

National Planning Department (DNP): National Plan of Development (2007).

National Department of Science and Technology: Special Project "Networks of Intellectual Property Management and Capacity Building Research Innovation And Business Development Direction. 2007.

WIPO: Research Networks and Intellectual Property: A Model for Supporting Developing Country Researchers in Creating, Protecting, and Exploiting Research Results in the Field of Health. (2006).

Section III

Asian Continent

Chapter 14
Intellectual Property Rights in Iraq

Teriza Jasim Ridha
Director of IPO,
COSQC
Baghdad - Iraq
E-mail: traiza_ipo97@yahoo.com

ABSTRACT

Intellectual Property Rights in Iraq deals with what the human thought innovates in the fields of patent, agricultural, industrial and scientific development as well as undisclosed information, cultural and artistic works, trademarks, service and commercial symbols and slogans and the names used in trade.

Commenting the orders of the Coalition authority, the council of ministers called the COSQC in 2005to prepare a law for Intellectual Property in Iraq taking into consideration linking IP rights in state department to be consistent with international law and in conformity with the entry of Iraq to the World Trade Organization and appropriate to the philosophy of the state's economy.

The COSQC acts with the collaboration of inventors and innovators sponsorship committees in all sectors to get benefit from the registered patents and invest them

Keywords: *Intellectual Property Rights, Iraq, COSQC, Ministry of Planning.*

Introduction

Intellectual Property Rights in Iraq deals with what the human thought innovates in the fields of patent, agricultural, industrial and scientific development as well as undisclosed information, cultural and artistic works, trademarks, service and commercial symbols and slogans and the names used in trade.

Intellectual Property (IP) is divided into two parts:

Industrial Property

Which includes patents, drawing and industrial designs, trademarks, geographical indications and trade names.

Copyrights

Includes literary specifications: novels, poems poetry, plays and films, and the artistic specifications: such as drawing, oil paintings, photographs, sculptures and architectural designs.

The IP rights like other property rights, allow the innovator or the owner of patent, or trade mark or copyright to benefit of his work and investment, under article 97 of the International Announcement of Human Rights which states on the right to benefit of the financial and moral interests protection of the proportion of output resulting from scientific, literary or artistic to the author.

Distribution of IP Rights

Industrial Property Division at the Central Organization for Standardization and Quality Control/Ministry of Planning

Previously, the said division was dealing with the four industrial property fields protection (Patents, Industrial designs, Trademarks and Trade names) this function continued till the eighties of the last century, when trademarks inflict with Iraqi Industries Union and trade names with Trade Chambers Union (non-gov.org.) to deal recently with protection of Patents and Industrial Designs fields.

Industrial Property Division Constitute of

A: Patents and Industrial Designs Section.

B: Investment Section.

C: Innovative Works Documentation Section.

The adopted law is Patent and Industrial Design Law No. 65 of 1970 (amended).

Patents

Each new innovative idea capable of industrial application reached by the inventor which is related to a new industrial product, or a method or created means or both together, leads to solve a certain problem in any of the technical areas.

Patent is granted also to a person who developed or improved the well-known Industrial Method which led to production increase or quality improvement.

The duration of Patent protection is 20 years as of the date of filling.

Industrial Drawing

Each installation or coordinate lines lending on the product elegance or earn a special form, whether it is using machine or manually, including textile designs.

Industrial Design

Each embodied form, whether associated with lines or colors or not, and given a special appearance can be used for Industrial or Craft Purpose.

Duration of protection of the two above is 10 years as of the date of grant.

The system used in Iraq for invention examination and evaluation is the previous exam system depends on existing of the 3 elements: novel, inventive and industrially applicable. And this will be done through assignment the applications to recognized specialist examiners and registered as evaluators as well as offering before consultant committees to recommend on the acceptance or rejection.

All patents and industrial designs applications will be documented after obtaining certificate under certain regulations also circulation will be made for the patent investment via calling for the beneficiary's body for using or investment.

Number of patents granted till now is:

- ✰ 1291 Patents for Citizens.
- ✰ 2024 Patents for foreigners

Also number of Industrial designs granted is:

- ✰ 625 Industrial designs for citizens.
- ✰ 152 Industrial designs for foreigners

The number of foreign countries eligible for registration is about 50 foreign countries.

Trade Marks and Particulars Division/Ministry of Industry and Minerals

Previously, Trademarks division was linked to Iraqi Industries Union (non-gov.org) but it is now linked with the ministry of industry and minerals.

The adopted Law is No. 21 of 1957 for trademarks and Geographical indications is adopted.

This division tasks is protecting the following marks:

Trade Mark

Each signal or group of signals may constitute a trade mark by which distinguish items of a project for goods of other projects such as signals and in particular words including personal names, letters, numbers, shapes, colours and symbolism as well as any combination of these signals can be registered as a trade mark. And these signals enable to distinguish the same goods and services the possibility for registration depends on the distinguished characteristic gained from the use, signal is not required to be visually perceived for protection as trade mark.

Service Mark

Each signal or group of signals used by a person to identify and distinguish services including individual service from others services and in order to refer the services sources even if it isn't well known.

Personal names and addresses and distinguished items of radio and Television programs may be registered as service mark even though these programs were announced for the sponsor goods and will be protected as service mark of retail mark whether for goods sale of the services provider to those owned by others or both.

Guarantee Mark

Each signal or group of signals used by a person not the real owner to state the regional origin or any other origin, material or manufacturing method or quality or accuracy or any other properties of the goods or services or that making product or services has been done by members of union or other organization include the guarantee mark of Geographical Indication.

Geographical Indication

References specified a good origin in region of state or territory or place in that region when the quality, feature or other properties of the good refers mainly to the geographical origin.

Collective Mark

It is a trade or service mark used by the members of collaboration or union or any group or collective organization contain marks refer to the membership in a union or association or any other organization.

The number of registered mark up to now is 65800 where the ratio was 30 per cent for citizen and 70 per cent for foreigners.

Trade Names/Iraqi Trade Chambers Union

Trade name is nomination or distinguished signal by which a corporation will be designated.

The trade names will be protected through registration the trade name of moral persons or associated or mixed companies under the provision of trade law No.30 of 1984 and trade names regulation No.6 of 1985 through approaching the merchant or company's owner to the trade chamber for checking that no similar name is exist then move to the companies registrar at the ministry of trade for the purpose of issuing an establishment certificate recognized by the state offices.

The total number of trade names registered up to now is (1212) consist of registration of Mix, Simple and Private Companies.

National Centre for Protecting Copyright and Neighbouring Rights (Ministry of Culture)

Previously copyright section was linked to the ministry of culture and information and the adopted copyright law is 3 of 1971 legislated and not applicable, where the protection was limited on protecting authors in publication and translation house. Then national centre has been recently established to begin applying the protection for each product in the field of scientific, artistic and literary work to include all audio-visual classifications.

Coalition Provisional Authority: Orders of 2004

To express the desire to create an important changes at the Iraqi intellectual property system rules as necessary to raise the Iraqi people economic level.

To improve living condition and technical skills and to create work opportunities in order to reduce unemployment and its attendant detrimental effect on the general security.

In recognition of the need of companies, lenders businessmen for a fair, effective and well known environment to protect their intellectual property.

In recognition of the desire that Iraq will be a full member in World Trade Organization.

In line with the report of Secretary-General of the UN Council on July 17, 2003 in respect of growth and development of Iraq to convert from the central economic planning to a free Market economy by establishing a mechanism for the private sector and the need for Institutional work.

It Declares Herewith

1. Order No. 80

   Law Trademark and Geographical Indication
2. Order No. 81

   Law of patents, Industrial designs, undisclosed information, Integrated circuits and plant varieties
3. Order No.82

   Law of Copyright

Law of IP in Iraq

Commenting the orders of the Coalition authority, the council of ministers called the COSQC in 2005 to prepare a law for Intellectual Property in Iraq taking into consideration linking IP rights in state department to be consistent with international law and in conformity with the entry of Iraq to the World Trade Organization (WTO) and appropriate to the philosophy of the state's economy. Therefore, some main and subsidiary committees have been formed and include the following bodies which are concerned with IP rights:

- ☆ The Central Organization for Standardization and Quality Control/ Ministry of Planning
- ☆ Ministry of Health
- ☆ Ministry of Industry and Minerals
- ☆ Ministry of Trade
- ☆ Ministry of Culture
- ☆ Iraqi Industries Union

- ✩ Chambers of Commerce Union
- ✩ USAID/WTO project

These bodies have unified and amended the above mentioned laws to conform to what are stated in the Agreement on Trade Related Aspects of Intellectual Property Rights, TRIPS, during the period from 2005 to 2010 to reach to integrated project satisfied to all requirements and the said project is under negotiation at the Consultancy Council of the state to pave the way for legislation.

The Role of COSQC and IP Rights

1. The COSQC is considered as Iraq representative and the permanent member in WIPO as well as Paris Convention for Protection of Industrial property since 1976.
2. The COSQC is a liaison point between World Intellectual Property Organization (WIPO) and the third party participating in IP activities in IRAQ.
3. The COSQC seeks for joining Iraq to the following agreements and conventions related to WIPO which are under legislation in Council of Ministers:
   - A. Patent Cooperation Treaty (PCT).
   - B. Madrid Agreement and its Protocol concerning the International Registration of Marks.
   - C. Hague Agreement concerning the International Registration of Industrial Designs.

1. The COSQC acts as mediator to circulate for Iraqi inventions and innovations' investment, through holding meetings between the inventors and the beneficiary bodies. The number of patents registered investments amount to 113.
2. The COSQC seek to conduct a Memorandum of Understanding with WIPO to get benefit of its development projects, also to join to other agreements of WIPO's, to get benefit of their consultant experiences related to prepare data bases and auditing IP draft law of Iraq the new.
3. The COSQC seeks for encourage inventors and circulate their innovations through national exhibitions and participation at international ones, such as international exhibition for invention and innovation which was held in Syria in 2009, in which 42 Iraqi inventors and innovators were participated and gained 16 Golden, Silver and Bronze prizes with WIPO's gold medal for the best woman inventor .
4. The COSQC is seeking to conduct MOU with brother countries as well as other developed countries in the field of IP right for the purpose of transfer technology and experiences development.

5. The COSQC acts with collaboration of USAID/Tatweer (National Capacity and Development Administration) to prepare regulations and data bases for inventions and innovations documentation.

The COSQC acts with the collaboration of inventors and innovators sponsorship committee, newly formed by the Ministry of Science & Technology, to circulate for inventions investment and support the illustrative models of the innovations that speed in inventions evaluation and examination.

Chapter 15

Intellectual Property Rights in Pakistan: Role of PCSIR for Strengthening the IPR System for National Growth

Syeda Rumla Naqvi[1], Nighat Afza[2] and Tanzil Haider Usmani[3]

[1]Industrial Liaison Officer; [2]Chief Scientific Officer; [3]Director General
Pakistan Council of Scientific and Industrial Research,
Shahrah-e-Dr. Salimmuzaman Siddique, off University Road Karachi, Pakistan
[1]E-mail: greatnaqvis@hotmail.com
[2]E-mail: hocprc@yahoo.com
[3]E-mail: klcpcsir@hotmail.com

ABSTRACT

Intellectual Property is an important factor in worldwide economic growth and in all knowledge and technology based industries and businesses. This is observed especially in the economic development achieved by some countries in the 1990's.The new concept of IP based nation is gaining ground because it is Intellectual Property which enables technology creation and technology transfer by providing the necessary enabling environment.

For these considerations Intellectual Property was mainstreamed in Pakistan in 2005. IP was brought under the limelight of public policy by simultaneously establishing IPO-Pakistan, empowering the Federal Investigation Authority (FIA) and activating Pakistan Customs against IPR violations. The Government decision to establish IPO-Pakistan was particularly aimed at addressing the institutional shortcomings that were impeding the effective management of intellectual property in Pakistan. IPO-Pakistan is now fast developing into a model of excellence in integrated management of IP Registries administering Patents, Trade Marks and Copyrights.

In Pakistan it will be an uphill task to introduce Patent culture of innovation. However for the diversion on commercial side they should devote their major portion of skills in developing commercially viable products for the society. The awareness programmes about Intellectual Property Rights will impact on our people mind set. PCSIR accept these challenges because it has already substantial know-how as regards the value and weightage of Patent in the WTO Scenario and also about the difficulties being faced for filing of Patent in Pakistan due to the unavailability of data base and complicated legal aspects. In the case of international Patents this scenario is rather more complicated. PCSIR Patent Advisory Cell is providing relevant consultancy to the Scientists/Technologists in all legal aspects for filing of Patents and provides the searching facilities and links for searching new technologies/processes and patents all over the world.

Keywords: *Intellectual Property Rights (IPR), Pakistan, Intellectual Property Organization of Pakistan (IPO), Pakistan Council of Scientific and Industrial Research (PCSIR), Patent Advisory Cell (PAC), Technology Business Incubation Centre (TBIC),Technology Transfer for commercialization of Patents.*

Introduction

The protection of property rights, including intellectual property rights (IPR), is a central pillar of a business-friendly governance structure in modern economic development. An efficient, transparent and rule-driven IPR regime in many countries has been considered instrumental in the innovation-led growth during the twentieth century. Subsequently, the emergence of an international trade regime under the World Trade Organization (WTO) has also been accompanied by a stricter Trade Related Intellectual Property Rights (TRIPS) regime. Just like several other countries, Pakistan has also undergone several stages of reforms in the governance structure and enforcement process of IPR since 2000.

General Information on Participant's Country

Pakistan, created in 1947, is a developing country of South Asia, spread over an area of 87.98 million hectares. The country is situated between latitude 23° and 37° north and longitude 61° and 76° east. The annual rainfall ranges from 125 mm in the South to 875 mm in sub-mountainous and northern plains. About 70 per cent of the rain falls during the monsoon season (July-September). However, occasional showers also occur during the winter. The summer months, except in mountainous areas, are very hot, while the winter months are mild in the plains and extremely severe in hilly regions. Capital of the country is Islamabad; major languages are Urdu (official),

Punjabi, Sindhi, Pashto, Balochi and English, life expectancies is 61 years for both men and women (UN). Main exports are textile products, rice, and cotton and leather goods. Monetary unit of Pakistan is Pakistani Rupee equivalent to 100 paisa. Per Capita Income is $1046 (Rs.80910.0). The state-owned forest area, under the control of the Forest Department, is 4.58 million hectares, only 5.2 per cent of the total area of Pakistan. Historical background of Pakistan is related to India and Iran on one side and to Arab countries on the other side. According to the latest census (2007 census) Pakistan is the ninth most densely populated country in the world. The population of the country is over 169 million, sex ratio is 1.05 male per female (nearly 49 per cent female) and life expectancy is 63.9 years. 97 per cent of the population is Muslims while according to 2009 World Population data Sheet the population of the country is 180.8 million, the sixth most densely populated country of the world.

Those who can read and write and are above 15 years of age are 48.7 per cent of the total population; male literacy rate is 61.7 per cent while that of female is 35.2 per cent.

Intellectual Property Rights' (IPR) Protections In Pakistan

Pakistan is a member of the World Trade Organization (WTO) and is also a signatory to the Trade Related Intellectual Property Rights Agreement, commonly known as the TRIPs Agreement, as well as a member of the Paris Convention. The regime of IP laws in Pakistan is well established and goes back to the early 20th century. Pakistan became a sovereign state after gaining independence from the British Empire in 1947. By virtue of this British rule stretching back to the mid-19th century, most of the laws in British India (comprising the territories now forming Pakistan, India, Bangladesh, Sri Lanka and Myanmar) were generally derived from the laws of the United Kingdom. When Pakistan gained independence in 1947, the intellectual property laws that were in force in British India at that time continued to remain in force in Pakistan. Among these laws, the main statutes included the Patents and Designs Act 1911, the Trade Marks Act 1940 and the Copyright Act 1914. In addition to these, the common laws for protection provided to intellectual property rights continue to apply.

Over the years, the above principal statutes have been amended to give way to new legislation. The Copyright Act 1914 was replaced by the Copyright Ordinance 1962. In the last decade or so, the laws have been amended in order to keep pace with the changing times and, in particular, major amendments have been introduced to bring Pakistani IP laws into conformity with the TRIPs Agreement. In this regard, for patents the act of 1911 was replaced by the Patents Ordinance 2000 and the Registered Designs Ordinance 2000. Also a Copyright (Amendment) Ordinance 2000 was introduced amending some provisions of the previous Ordinance of 1962. In the field of trademarks, a Trade Marks Ordinance 2001 was enforced in 2004.

Pakistan's International Commitments

The U.S. Pakistan Treaty of Friendship, Commerce, and Navigation guarantees national treatment for patent, trademarks and industrial property rights. Pakistan is a member of the World Intellectual Property Organization (WIPO), the Universal Copyright Convention, and the Bern Copyright Union. The United States and Pakistan have held a series of official discussions regarding intellectual property protection, preserving the rights of foreign companies and individuals, and Pakistan's active commitment to the Trade-Related aspects of Intellectual Property Rights (TRIPS). Pakistan became a signatory to TRIPS in 1995; as a result, the scope of foreign investment through the field of intellectual property rights has increased considerably.

Infringement of IP Rights in Pakistan

As is the case with most third world countries (and even some developed countries), intellectual property infringement is unfortunately very widespread in Pakistan, extending to all businesses and trades. Infringement of trademarks is mostly experienced in products from the apparel, electronics, food and confectionery, pharmaceuticals, and cosmetics industries. Importantly, the infringers in many cases adopt trademarks which are different; however, they produce look-alikes of the colour scheme, design and get-up of the product such as its packaging and trade dress. In this way, infringers attempt to tap the lower end of the consumer market in the country, which is not very well educated and is thus deceived by the look-alike packaging. With the improvement in packaging technology, and with more and more sophisticated procedures available to the infringers, it is becoming increasingly difficult for consumers to detect the counterfeits.

In the case of patents, as Pakistan has a large and upcoming pharmaceutical industry, there are a number of infringements relating to patented processes for pharmaceuticals, as well as patented compounds and compositions, by local generic manufacturers, who mostly import the generic drugs or their raw materials from China, Brazil and India, and repack and sell them in Pakistan. Patent infringements in other industrial sectors, such as engineering, are few and far between. Infringement of designs is also encountered in the engineering, electronics, textiles, food and beverages, and various manufacturing sectors. Fortunately, the legal system in the country provides for adequate protection for trademark, patent and design owners, and those owners who pursue the legal remedies aggressively and proactively are able to control the menace of infringement by due process of law.

International Rank of Pakistan

According to International Property Right Index (IPRI) report 2009 by Property Rights Alliance With regard to Pakistan, Intellectual Property Rights continue to be a challenge. This year Pakistan's score on the scale of IPRI decreased 0.6 points The Copyright Piracy score increased, but this change was only because there were no data for piracy of records and music in 2010. From 2007 to 2009, however, it is estimated that piracy accounted for the entire music market in Pakistan.

Pakistan-IP Situation before 2005

Before 2005, IP situation in Pakistan was being internationally perceived to be constantly deteriorating. Pakistan was being blamed for export of pirated optical discs to a number of countries of the world. The deeply deteriorating IP situation was also considered within the country to be highly detrimental to the following:

- ✰ Rapid growth mode of Pakistan economy;
- ✰ Enabling environment for investment attraction;
- ✰ Market access requirement of Pakistan's export-led growth strategy;
- ✰ Fast track integration with global economy;
- ✰ Market expansion for genuine businesses in Pakistan;
- ✰ Consumer interest in terms of competitive quality and price;
- ✰ Country's business image as a whole.

IP management in Pakistan could not keep pace with the fast changing global environment. Internationally, Pakistan was being blamed for export of piracy and counterfeiting. Government was perceived to be doing nothing to improve the situation. Continuation of this situation was, therefore, obviously not conducive to the investment and export interests of the country which are the key for sustaining and strengthening the growth of the economy.

Out-of-the box Solution

In order to redress the situation, Government of Pakistan found an out-of-the box solution for its IP problems and took the following three parallel decisions on the 8th April, 2005; namely:

- ✰ Establishment of IPO-Pakistan as a focal organization for integrated management of intellectual property and enforcement coordination;
- ✰ Empowerment of FIA (FIA stands for Federal Investigation Agency which is Pakistan's elite force for control of white collar crime) to eliminate piracy by including the Copyrights Ordinance 1962 (as amended in 2000) in the FIA Act, 1974;
- ✰ Activation of Pakistan Customs to effectively interdict import and export of pirated optical discs etc.

Intellectual Property Organization (IPO)-Pakistan

Currently, the IPR regime is governed by an integrated piece of legislation, called the Intellectual Property Organization (IPO) of Pakistan Ordinance 2005. The earlier regime was fragmented into Ministry of Commerce (for trademarks), Ministry of Education (copyrights) and Ministry of Industries and Production (for patents and designs) and now it has come under the umbrella of an integrated IP office in Islamabad, called the Intellectual Property Organization, attached to the Cabinet Division and directly reporting to the Prime Minister.

Present Situation

Today, IPO-Pakistan is one of the most integrated IP organizations of the world. All forms of IP have been integrated under one umbrella organization. The automation of IP Registries in Karachi and IPO Regional Office Lahore have been completed. The data capturing of old record is underway.

Core Functions

IPO- Pakistan has been assigned the following core functions namely:

- Integrating IP management;
- Improving service delivery;
- Increasing public awareness; and
- Enhancing enforcement coordination.

Governance Structure

The IPO-Policy Board effectively symbolizes the spirit of Public-Private Partnership which, in fact, is the new paradigm of participative governance in Pakistan. It also combines the finest expertise of public administration and corporate management in the best national interest. The Deputy Chairman, Planning Commission and six Federal Secretaries of the stake-holding Ministries represent the Public Sector and eleven Members including the Chairman represent the Private Sector including top executives of the national and multinational companies in Pakistan. Chairman Policy Board is a distinguished and widely experienced professional from the Private Sector. The Director General is the Chief Executive Officer (CEO) of IPO-Pakistan and Secretary to the IPO Policy Board. He is also a Federal Secretary (BS-22) which is the highest position in the civil service. Thus the Government of Pakistan has invested its finest human capital in the governance structure of IPO-Pakistan.

Registries Data of IPO-Pakistan

Table 15.1 shows the application received in IPO-Pakistan regarding registration of Trademark, Copyright and filing of Patents.

Table 15.1: Applications Received

Sl.No.	*Registry*	*2004-05*	*2005-06*	*2006-07*	*2007-08*	*2008-09*	*Increase/ Decrease % (+/–)*
1.	**Trademark Registry, Karachi**	13097	13129	14309	14170	16056	13.30
2.	**Copyright Office, Karachi**	1631	2762	2638	2346	3268	39.30
3.	**Patent Office, Karachi**	493	1406	1790	1535	1365	(-)11.07
4.	**Design Office, Karachi**	2479	652	487	535	389	(-)27.28
	Grand Total	**117700**	**17949**	**19224**	**18586**	**21078**	**13.40**

Source: Annual Report of IPO-Pakistan 2008-09.

Table 15.2 shows the number of application examined in IPO-Pakistan regarding registration of Trademark, Copyright and filing of Patents.

Table 15.2: Applications Examined

Sl.No.	*Registry*	*2004-05*	*2005-06*	*2006-07*	*2007-08*	*2008-09*	*Increase/ Decrease % (+/–)*
1.	**Trademark Registry, Karachi**	13411	16083	18857	14802	13092	(-)11.55
2.	**Copyright Office, Karachi**	1217	2486	3486	3098	3657	18.04
3.	**Patent Office, Karachi**	493	652	487	498	350	(-)29.71
4.	**Design Office, Karachi**	1635	1849	1623	2130	2687	26.15
5	**Grand Total**	**16756**	**21070**	**24453**	**20528**	**19786**	**(-)3.61**

Source: Annual Report of IPO-Pakistan 2008-09.

Table 15.3 shows the number of Registration Granted in IPO-Pakistan regarding Trademark, Copyright and filing of Patents.

Table 15.3: Registration Granted

Sl.No.	*Registry*	*2004-05*	*2005-06*	*2006-07*	*2007-08*	*2008-09*	*Increase/ Decrease % (+/–)*
1.	**Trademark Registry, Karachi**	3586	2833	5067	7412	5656	(-)23.69
2.	**Copyright Office, Karachi**	414	1247	1237	1500	2097	39.8
3.	**Patent Office, Karachi**	484	256	299	152	162	6.57
4.	**Design Office, Karachi**	872	468	397	438	359	(-)18.03
	Grand Total	**5356**	**4804**	**7000**	**9502**	**8274**	**(-)12.92**

Source: Annual Report of IPO-Pakistan 2008-09.

Public Awareness

IP awareness in Pakistan is very low. Many people did not get patent for their invention due to lack of awareness about rules and regulations of Pakistan Patent Office as well as for other countries. Due to this reason both economically and academically the benefits of novel patent or design is not being availed and a lot of innovative talent of our country has been wasted. A lot of problems have been arisen to the individuals in obtaining patent at their own levels.

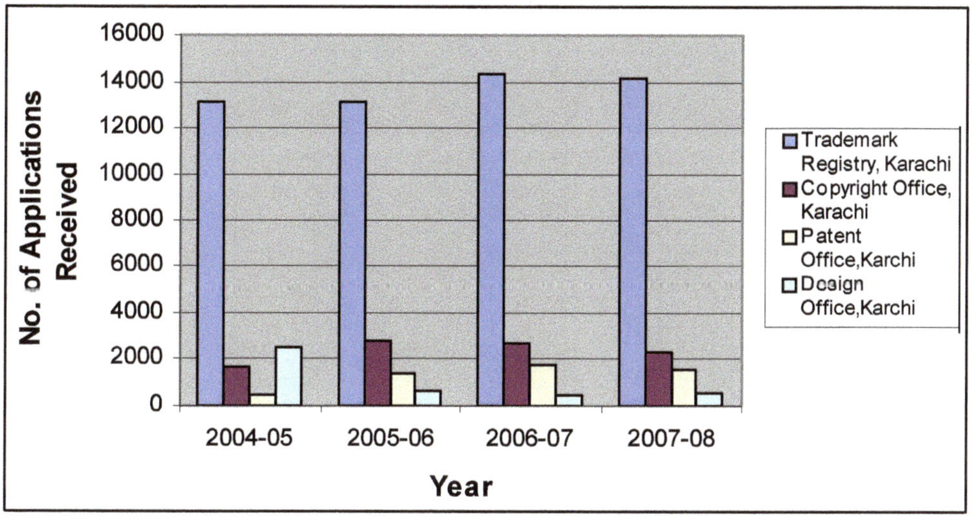

Figure 15.1: Graph of Applications Received in IPO-Pakistan.

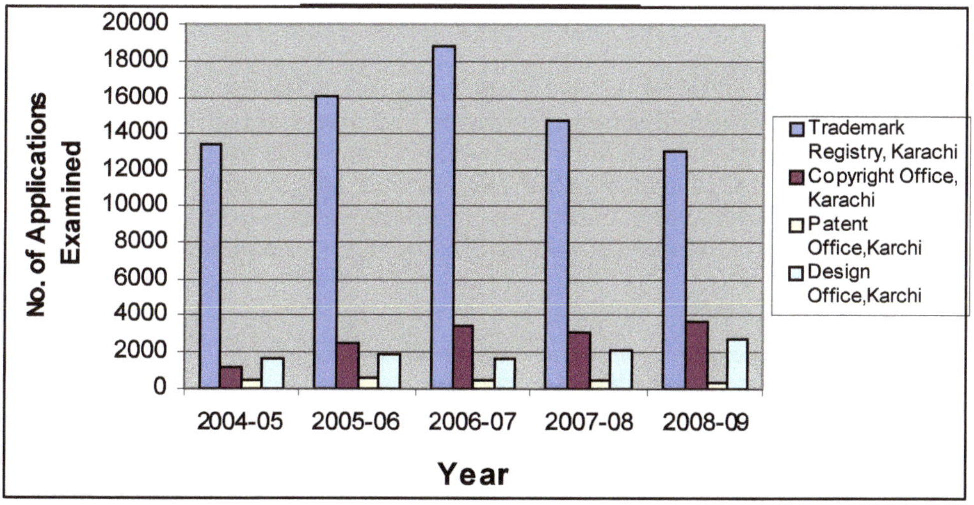

Figure 15.2: Graph of Applications Examined in IPO-Pakistan

General Information on Participant's Organization

Organization at a Glance

The Pakistan Council of Scientific and Industrial Research was created in 1949 as a department of Ministry of Industries. It is a premier R&D organization, the largest in the country, for scientific and industrial research and has been involved in multi-disciplinary activities for more then half a century. PCSIR was reconstituted as an autonomous Council in 1953 and was placed under administrative cannot of MoST in 1972. The main function of the Council is to undertake Research and Development efforts directed towards utilization of indigenous resources for

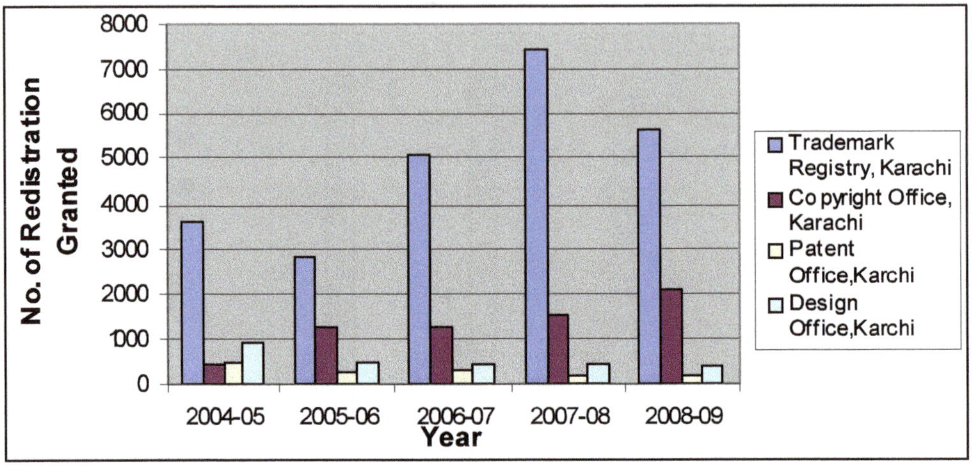

Figure 15.3: Graph of Registration Granted Through IPO-Pakistan

industrial development and import substitution leading to self-reliance. PCSIR has established multi-disciplinary laboratories in various parts of the country.

Organizational Set-up

Since 1953 PCSIR scientists have developed 1362 processes registered 465 patents locally and 29 foreign patents, published over 6700 papers in the fields of minerals, glass and ceramics, pharmaceuticals, electronics and instrumentation, agro-technology, food technology, industrial and fine chemicals, leather, fuel, oils and fats. PCSIR now has the capability of setting up various small and medium-scale

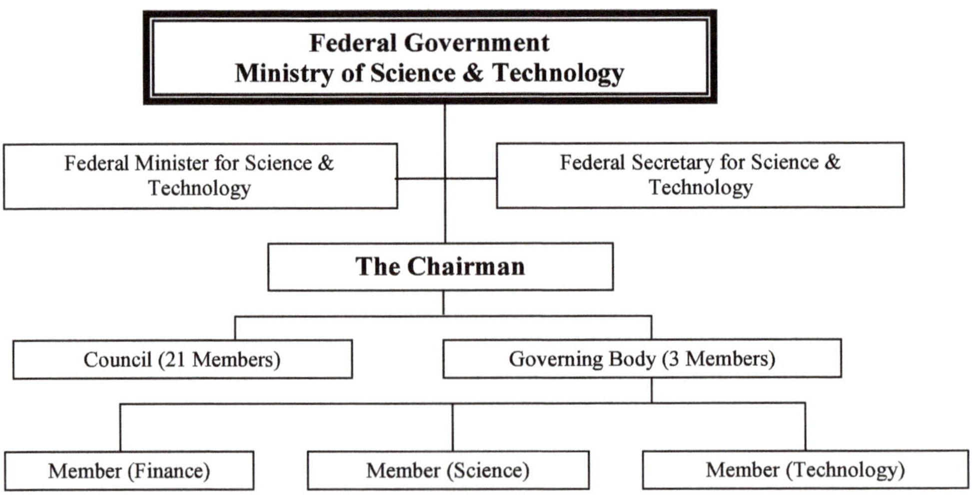

Figure 15.4: Organizational Set-Up of PCSIR

units for production. So far, 1127 processes have been leased out to various private and public sector organization.

Twelve PCSIR Laboratories currently are located in Karachi, Lahore, Peshawar, Islamabad, Hyderabad, Quetta and Skardu. Their activities include industrial chemicals, pharmaceuticals, leather technology, coal preparation and carbonization, food technology, minerals and metallurgy, glass and ceramics, water purification, drugs, rural technologies, solar energy plants, fruits and marine food.

Total working strength of PCSIR is 2408 out of which 850 are Scientists/ Technologists and 40 per cent of them are female, technical/skilled workers are 793 while administrative staff is 765.

Patent Culture in PCSIR

Patent culture in PCSIR has been nurturing since 1953 and has more than 500 patents, and has published more than 7000 research papers in national and international journals. It has identified potential tangible and intangible assets that could be commercialized and produce commercially viable products. To encourage and facilitate such innovations, various funding schemes are required for necessary research and development by the government. He pleaded for the need to continue creation of new inventions which are essential for the development of nation.

Technology Business Incubation Centre (TBIC): (A Way to Commercialization of Patents)

Innovation is of key importance in spurring economic growth in a developing country. To achieve the level of developed countries have to focus on every field of science and technology; do research to solve the problems of nation.

Technological development is the most reliable tool for economic stability of a nation; the scientific and technological research must be translated into commercially viable processes and products, which may be utilized by the industry.

Among the most important futuristic initiatives for the commercialization of techniques, PCSIR has taken a bold decision of setting up a variant version of Technology Business Incubators (TBIC). Through these TBICs, PCSIR is aiming at popularization of its technologies in an innovation approach and promote the SMEs. These TBIC will function as pilot plants for the demonstration of technological processes developed by its scientist.It provides entrepreneurs not only flexible terms, but also the infrastructure and business support services they initially need in order to be successful.

These Centres contribute to the economy of the country in the following ways:

- Create employment for rural and urban workforce.
- Provide flexibility and innovation in the economy.
- Alleviation of Poverty.
- Contribution to Exports.

Promoting Understanding of IP and Realizing its Development Potential:

In today's world, R and D organizations and universities can play a crucial role in the development of national economy, especially invention and innovation. We have to focus on every field of science and technology to match the developed countries.

To strengthen competitiveness in market, to work in sync with the technological advancement of Pakistan and world, to set up the IP protection system in conformity with the stages of national development, to foster public awareness of IP protection and to guide enterprises, universities and research institutions so as to improve their capacity to utilize, manage and protect IP with creating awareness and ensuring protection to the aware, PCSIR with the support of Ministry of Science and Technology has established the much needed Patent Advisory Cell. It is certainly a valuable helping hand for inventors and innovators regarding matters for legal protection of their achievements.

PCSIR Patent Advisory Cell (PAC)

Patent Advisory Cell has been established in a PSDP Project in 2006 in PCSIR Karachi, Pakistan funded by Ministry of Science and Technology. This cell encourages Pakistani Scientists/Technologists, Engineers and other stakeholders for filing National and International Patents of their newly developed technologies/processes and inventions. Patent related literature is easily available in Patent Advisory Cell library, searching facilities and all type of consultancies are also provided by this Cell.

Our Scientists, Technologists and Engineers should get awareness about the Intellectual Property Rights and Patent System, their importance and the expected financial benefits in term of royalty after commercialization of their patentable inventions. In Pakistan it will be an uphill task to introduce Patent culture of innovation. However for the diversion on commercial side they should devote their major portion of skills in developing commercially viable products for the society. The awareness programmes about Intellectual Property Rights will impact on our people mind set.

By the Cell total **225** national patents have been filed and **twenty six (26)** patents have been commercialized. **Seventeen (17)** foreign patents have been filed in USA, Japan, China, UAE, European Community, India and UK. **Six (06)** foreign patents are published in their respective countries.

Objectives of PAC

- ☆ To establish Patent Advisory Cell for Scientists, Engineers of S&T/R&D Organization and Institutions to acquire National and Foreign Patents for their Processes/Technologies/Inventions.
- ☆ To encourage the Scientists/Technologists to carry out scientific and technical research for filing patents of new technologies and processes.

This will be achieved through Seminars/Lectures and Techno-economic benefits in term of royalty and incentives etc.

- Patent Advisory Cell is providing technical back-up and assistance to the prospective patentees and providing all the relevant facilities like searching/consultancy/designing/drafting etc of a patent specification.
- To conduct Seminars and Workshops for creating awareness about IP Rights and its benefits (both socially and economically) for Socio economic prosperity of individuals, R&D organizations of public and private sectors and all the stakeholders.
- To provide technological information based on patent literature and documentations to various stakeholders like small and medium industries (SME's) and to the news and print media for its proper dissemination.

Proposed Facilities to R&D Organization/Institutions

- Patent search with patent applications at National and International Levels
- Efficient cost effective service.
- Technical Guidance and Consultation to obtain national and foreign patents.
- Quality work to client's satisfaction.
- Legal advisory services to ensure the best protection of inventions and innovations.
- To promote and support moves to simplify and modernize the law on intellectual property and the harmonization of international rules and procedures.
- Consequently services for grant of patents and registration of designs with a good presumption of validity and based on excellent standards of service, measured against benchmarks is set in consultation with users. Also provide service for maintenance of rights granted of registered, including patents granted by foreign patent office under the joint patent agencies like that PIPRO, WIPO, etc.
- To support PIPRO through consultancy.

Future Challenges

To put Pakistan on the IP map of the world as a compliant and responsible country by promoting and protecting Intellectual Property Rights (IPR).

Conclusion

In the present time the name of the game is K-Economy. The world now calls the present era as Innovation–Economy. It is need of today to establish National Innovation Centre in the country and it will be supported by the Ministry of Science and Technology. In today's modern world R and D organizations and Universities can play a crucial role in the development of national economy especially through invention and innovation

Innovation is of key importance in spurring economic growth in a developing country. To achieve the level of developed countries have to focus on every field of science and technology; do research to solve the problems of nation.

Technological development is the most reliable tool for economic stability of a nation; the scientific and technological research must be translated into commercially viable processes and products, which could be utilized by the industry. However, "This is only possible if the researchers focus their attention on patentable research".

IP is not an end in itself. It is a means towards an end. The end is sustaining and strengthening the ongoing high growth mode of the national economy and to ultimately achieve the goal of being an IP efficient society and an IP based nation. Therefore, strong IP regime is a paramount national imperative for achieving the objective of competitive economy based on:

- ✰ Technology Creation;
- ✰ Technology Transfer; and
- ✰ Creativity Promotion.

Acknowledgements

I am very much grateful to Mr. Zuzzer Ali Shamsuddin, Director, Leather Research Centre, PCSIR and Mrs.Askari Begum, Director, Planning and Development PCSIR Labs Complex, Karachi for their kindness and cooperation. They gave us great courage to face this challenge and make our work easier and comprehensive.

My thanks are due to my in-charge Mrs. Noor Taj, PSO-OIC/ILO and my colleagues for their cooperation during course of investigation.

References

Anne Chandima Dedigama, 2008 Hernando de Soto Fellow, International Property Right Index (IPRI) 2009 Report by Property Right Alliance.

Business Horizons, 2008. Pakistan's Intellectual Property Rights Regime: Investigating and Explaining Change.

http//www.mapsofworld.com/Pakistan/population-of-pakistan.html,Pakistan Population

Iqbal Mirza, March 03 2010, Daily Business Recorder, IPR role in economic growth: on need for budgetary support for R&D outfits stressed.

http://www.ipo.gov.pk,Intellectual Property Organization, Government of Pakistan, Annual Report 2007-08.

M.Farrukh Irfan Khan reports,Pakistan Risk Management,2007,Intellectual Property Laws in Pakistan http//www.mapsofworld.com/Pakistan/population-of-pakistan.html,Pakistan Population

http://www.propakistan.com,Pakistan Economic Survey 2008-09; International Forex rate.

http://www.prb.org/pdf09/09wpds_eng.pdf © 2009 Population reference Bureau,2009 World Population data Sheet

http://www.pseb.org.pk, Sajid Iqbal Manager Domestic Business, Pakistan Software Export Board (G) Limited, Intellectual Property Rights' (IPR) Protection in Pakistan.

Report on achievement of Pakistan Council of Scientific& Industrial Research (PCSIR), 2004.

Chapter 16

Status of Intellectual Property Rights (IPR) with Reference to Genetic Resources in Nepal

Chiranjivi Regmi

Ph.D., Chief, Faculty of Technology
Nepal Academy of Science and Technology,
Lalitpur, Nepal
E-mail: info@nast.org.np; cregmi@hotmail.com

ABSTRACT

Nepal is rich in biodiversity due to its vertical micro-climatic zones ranging from tropical to alpine. Nepal's share of worlds land is 0.1 per cent while its share in floral and faunal species is over 2 per cent. In Nepal 10107 floral and 6570 faunal species have been recorded, several of them are endemic to the country and many more are endemic to Himalaya.

Nepal signed on CBD on June 12, 1992 expressing the country's obligatory commitment of conserving its biodiversity. By doing so, Nepal is committed to contribute to achieve the objectives of CBD. Nepal, in order to protect its biodiversity including Indigenous Knowledge and Technology (IKT) has to develop a legal registration of its biodiversity by developing its own *sui generis* system so that its total biodiversity will be legally documented.

Nepal has acquired membership of World Trade Organization (WTO) in 2004 and thereby signatory of Agreement on Trade Related Aspects of Intellectual Property Rights (TRIPS), it has to make legal and organizational arrangements to meet the obligations under WTO.

The Ministry of Forest and Soil Conservation and the Ministry of Agriculture and Cooperation have been designated as the Focal points for dealing biodiversity of flora and fauna and agricultural biodiversity respect5ively. Accordingly Department of National Parks and Wildlife Conservation is making efforts for conservation and protection of biodiversity through Protected area system *i.e.*

in situ condition, Department of Plant Resources through live plant collection *i.e. ex situ* in Botanical Garden while Nepal Agriculture Research Council is doing it through Germ Plasm Collection and establishing Gene Bank.

Efforts on Protection of biodiversity and Indigenous Technology Knowledge thereof in relation to the Intellectual Property Rights have been initiated in Nepal but the advancement of implementing endeavor seems to be slower. All the efforts made by the Government of Nepal to handle global IPR issues specially related to protection of Genetic Resources and Traditional Knowledge are discussed in this chapter.

Keywords; *Genetic resources, Indigenous knowledge and technology, Biodiversity, Micro-climatic zones, Intellectual property rights, Ethnic groups, Breeders right.*

Introduction

Nepal is rich in biodiversity due to its vertical micro-climatic zones ranging from tropical to alpine attributed by the altitudinal variation of the topography. As a result 75 types of vegetation including 35 types of forest have been recorded within the comparatively small territory of the country. The upper one (above 3000m asl) has floristic affinity to Central Asiatic floristic region where as Upper subtropical and temperate bio-climatic belt (1550-3000m asl) is related to Sino -Japanese floristic region and the area below 1500 m is related to Indian and Southeast Asia -Malaysian Floristic region. These factors have created very rich biodiversity that has contributed to rank Nepal between 25th and 30th position in global context and 11th in the continental scale of biodiversity richness. Nepal covers 0.1 per cent of the world land but shares 2.5 per cent of the global flora.

Nepal has floral wealth of 10107 species (2.5 per cent of the world's total) of which 5891 are flowering plants. Of the flowering flora 5 per cent are endemic to the country and 30 per cent endemic to the Himalayas. Among the non flowering plants 387 species of Pteridophytes, 471 lichens, 853 bryophytes, 1822 fungi and 687 algae have been recorded in the country. Similarly, 6570 species of fauna have been recorded 186 species of mammals and 847 of birds representing 4.3 per cent and 8.5 per cent of the world occupying 10th and 6th positions in the list of the Asian countries which are richest in mammals and bird respectively.

Agricultural biodiversity of Nepal is also very rich. This is apparent from the 172 families, 294 genera and 551 species/subspecies of agricultural crops. Of these, 200 species are cultivated. Crops such as rice, buckwheat, citrus, banana and mango have high genetic diversity as the center of origin of these crops falls in this region. Similarly, the diversity of under-utilized crops and tropical and subtropical fruit species is noteworthy. About 215 improved varieties of 44 crops have release so far in the country. Out of them 35 varieties have become out dated and are not recommended for cultivation any more. In fact only 7 per cent of the farmers are using seeds made available from the authentic sources and the rest 93 per cent of them are using the seeds stored by them, produced by them, exchanged among each other or procured from unofficial sources.

In fact Nepal has its own type of intellectual heritage of preserving its genetic diversity. For example, it has maintained Rice-Jumli Marshi that has been cultivated in temperate climatic condition in Jumla and Bajura area in west Nepal while Ghaiya dhan that are cultivated in on well drained sloppy area in suburban areas of Kathmandu. Similarly cultivation of Uva (*Hordium vulgare)* at high altitude up to around free line 4000 m in east Nepal. The farmers are using the indigenous traditional knowledge for food security by using traditional technologies developed by their ancestors. These farmers are unknowingly preserving their intellectual heritage of preserving genetic materials of the upland rice and Himalayan barley, which might be of immense value for plant improvement.

There are about 60 ethnic groups residing in Nepal. Their undocumented traditional knowledge on utilization of plant materials is the intellectual heritage of Nepal. Contribution of intellectual heritage in the form of indigenous/ethnic knowledge on biodiversity cannot be ignored during biodiversity utilization and conservation. Such ethno-botanical knowledge may become a precious tool to open up incurring dead locks in the advancement of some sectors like human health where there is an urgent need of life saving drugs.

Convention on Biological Diversity (CBD)

Convention of Biodiversity (CBD) was held at Rio de Jenerio, Brazil in 1992, which was participated and signed by 170 countries. Negotiated under the auspices of the United Nations environment Programme (UNEP), the CBD opened for signature on 5 June, 1992 at the Rio Earth Summit, and entered into force on 29 December, 1993. The Convention is legally binding and Contracting Parties are obliged to implement its provisions. Nepal signed on CBD on June 12, 1992 expressing the country's bold expression of an obligatory commitment of conserving its biodiversity. So far, 190 countries and the European Union are its members.

The objectives of CBD are; conservation of biodiversity, sustainable use of its components and the fair and equitable sharing of benefits out of the use of genetic resources by appropriate access to genetic resources, transfer of relevant technologies and funding.

Article 3 of CBD mentions " States have sovereign right to exploit their own resources pursuant to their own environmental policies and the responsibility to ensure that activities with in their jurisdiction or control do not cause damage to the environment of other states or of areas beyond the limits of national jurisdiction".

Article 15 of the CBD provides a framework for the implementation of ABS. In recognition of the sovereign rights of states over the biological resources, national governments, subject to their national laws, are conferred the authority to determine access to genetics resources. The Convention requires its parties to create conditions, subject to allow safeguard, to facilitate access to genetic resources for environmentally sound uses by other parties on a bilateral basis.

World Trade Organization and Intellectual Property Protection

The World Trade Organization was established in January, 1995 during the eighth Round Talk of GATT (General Agreement of Tariff and Trade) in Uruguay. WTO concentrates on improvement of global trading environment which mainly covers abolition of tariff and non tariff barriers to trade, greater transparency on trade practices, clear cut procedure for settlement of disputes, protection of intellectual property rights and opening up of service sectors. Trade in goods, trade in services and TRIPS are the three main components of WTO.

Trade Related Aspect of Intellectual Property Rights (TRIPs)

TRIPs in the Agreement for world trade provides that the protection should be extended to Copyright and related rights, Trademarks, Geographical Indicators, Industrial Design, Patents, Lay-out Designs (Topographies) of Integrated Circuits, Trade Secrets and the protection of plant varieties in the national level for which, patent or *sui-generis* system may be introduced. In TRIPS Agreement, the patent protection has been extended to the all inventions whether product or process and while defining patent it specifies such invention which are new and involves inventive steps in it and that should be capable of industrial purposes. Its Articles 27 compels member countries to protect through patents innovations in all fields including food, health and other biotechnology-related fields. Member countries may however exclude plants, animals and essentially biological processes for reproduction from patenting. However, now it is mandatory to protect plant varieties through patents or any other independent *sui generis* system or combination thereof that are effective. In this respect, the member countries are forced to adjust their national systems for the protection of such life forms and related Indigenous Knowledge and Technology (IKT) to the rigid standards already established in the industrialized countries.

The International Treaty on Plant Genetics Resources for Food and Agriculture (ITPGRF)

The International Treaty on Plant Genetic Resources for Food and Agriculture (ITPGRF) entered into force on 29 June 2004. In addition to dealing with various aspects of conservation, management and sustainable use of Plant Genetic Resources for Food and Agriculture (PGRFA), this historic accord, while representing a legally binding, international commitment to improvements of world's key food and feed crops provides for farmers' rights. After accession to the Treaty, there is a need for each Contracting Party to review its existing policies, law and regulations and, if required, adopt new policies, law and regulations to ensure that the obligations under the Treaty are effectively implemented. In this regard, it is crucial for each Contracting Party to work out a strategic plan to implement the Treaty within the context of its national vision of social, economic and cultural development, and international cooperation.

Nepal's efforts in Protecting Intellectual Property Rights

Patent Design and Trade Mark Act came into existence for the first time in 1936 and repealed in 1966 in order to protect Intellectual Property Right. Patent Design and Trademark Act, 1966 is still in existence with amendments in 1987 and 1991. Similarly the Act on Copyright was promulgated in 1965, which is still prevailing. These two Acts protect the Intellectual Property Rights in Nepal.

Nepal signed on CBD on 12 June, 1992, ratified on 14 September, 1993 and it has already entered into force from February 1994. In 2002, Nepal developed a comprehensive Nepal Biodiversity Strategy (NBS) with the participation of a broad cross-section of Nepali society as well as in consultation with international experts to fulfill its obligations of being a party to CBD. This NBS is intended to serve as a guide to all government organizations, the private sector and civil society. As the part of implementation of NBS, the Government of Nepal (GoN) carried out extensive consultations with different stakeholders and experts and prioritized 13 concept projects for the period of 2006-2010 that comprises a cross-sectoral and six sectoral thematic areas such as protected areas, forests, rangelands, agriculture, wetlands and mountains; and are published in the Nepal Biodiversity Strategy Implementation Plan (NBSIP). Nepal has already prepared Fourth National Report to the CBD in 2009 encompassing the the progress made in implementation of NBSIP.

The Ministry of Forests and Soil Conservation has been designated as the focal point for dealing biodiversity of flora and fauna. The ministry has already prepared "Draft Act" on Access to the Genetic Resources and Equity sharing of the benefits arising from them". The Department of National Parks and Wildlife Conservation is making efforts for conservation and protection of biodiversity through Protected area system i.e. *in situ* condition. At present the protected area system covers about 18 per cent of the total land area of the country. The Department of Plant Resources is dealing with the collection and conservation of live plants in *ex situ* condition in the form of Botanical gardens and Herbal Orchards.

Considering the fact that above-mentioned Act by virtue of its nature implementing modality does not cover the rights of plant breeders and the farmers, the Government of Nepal has issued "Agricultural Biodiversity Policy –2063" through the Ministry of Agriculture and Cooperatives. To achieve the objectives of the policy the government has assigned the responsibility of developing necessary laws and regulations for the over all development of agriculture, food security and protection and improvement of agriculture biodiversity. Accordingly the Ministry has drafted a law for the protection of rights of plant breeders and farmers. The policy covers the rights on the genetic resources of crops and their conservation, justified distribution of benefits from them among the stakeholders to ensure the food security.

Nepal, as the member of WTO belonging to the Least Developed need not to develop legal and binding arrangements for plant breeders rights, it has to develop system for the protection of seed and related knowledge of farmers up to 2013. Nepal Agriculture Research Council is doing it through Germ Plasm Collection and establishing Gene Bank. Similarly as per the obligation to make and implement the Crop Variety Protection Act, Plant Breeders Right Act and Farmers Right on TK and plant breeding Act, have to be implemented after 2013. Therefore the government has to work on this line sincerely.

Conclusion

Being a rich country in genetic resources, Nepal has both great oppurtinities and challenges for the sustainable utilisation and conservation of these resources. As a signatory of CBD and WTO it has to be prepared for fulfilling the obligations by developing capabilities for the protection of Intellectual Property Rights and Indigenous Traditional knowledge on Genetic Resources. Attempts made by the Government of Nepal in this aspects are very positive but not adequate. Therefore it has become necessary to give due priority for the protection of Intellectual Property Rights related to genetic resources.

References

Acharya, B. P. (2000) Intellectual Property Rights in the context of Nepal. Proc.Nat.Workshop on IPR in the Nepalese context. NAST p.1- 4

Balmford, A., Bennun, L., ten Brink, B. *et al.* 2005. The Convention on Biological Diversity's 2010 Target. *Science,* vol. 307, Issue 5707, pp 212-213.

Belbase, N. 1999. National Implementation of the Convention on Biological Diversity: Policy and Legislative Requirements. IUCN Nepal, Kathmandu, Nepal.

Shakya, P.R. (2000) Intellectual Property Rights in relation to biodiversity. Proc.Nat.Workshop on IPR in the Nepalese context. NAST p.5-15.

Chapter 17

The Current Situation of Intellectual Property in the Kingdom of Cambodia

Chhean Piseth

Chief of Lay-out Design of Integrate Circuits and Licensing Contract Office, Ministry of Industry, Mines and Energy Department of Industrial Property, Cambodia

E-mail: chheanpiseth@yahoo.com

ABSTRACT

The Ministry of Industry, Mines and Energy of the Kingdom of Cambodia, is responsible for the protection of Patent, Utility Model, Industrial Design, Lay-out Design of Integrate Circuits and New Plant Variety. The Kingdom of Cambodia is combating and strengthening enforcement of the IP. People have little knowledge about Intellectual Property and there is lack of public awareness on intellectual property. The Ministry of Industry, Mines and Energy are upgrading the knowledge and skills for professional officials on the field of Intellectual Property by sending them to attend training courses, seminars and workshops in this field, locally or abroad, strengthening and widening the cooperation with other International Organizations and other Intellectual Property Offices in exchanging information and experiences for the development of Industrial Property System.

***Keywords**: Intellectual property, Kingdom of Cambodia, Industrial property system.*

Introduction

The Kingdom of Cambodia is situated in the Mekong sub-region, bordered by Thailand to the west and north, Laos to the north, Vietnam to the east and south-east of the Gulf of Thailand to the south-west. It has a land area of 181,035 square kilometres and the population is around 15 Million.

Administration of Intellectual Property System

The Ministry of Industry, Mines and Energy is responsible for the protection of Patent, Utility Model, Industrial Design, Lay-out Design of Integrate Circuits and New Plant Variety.

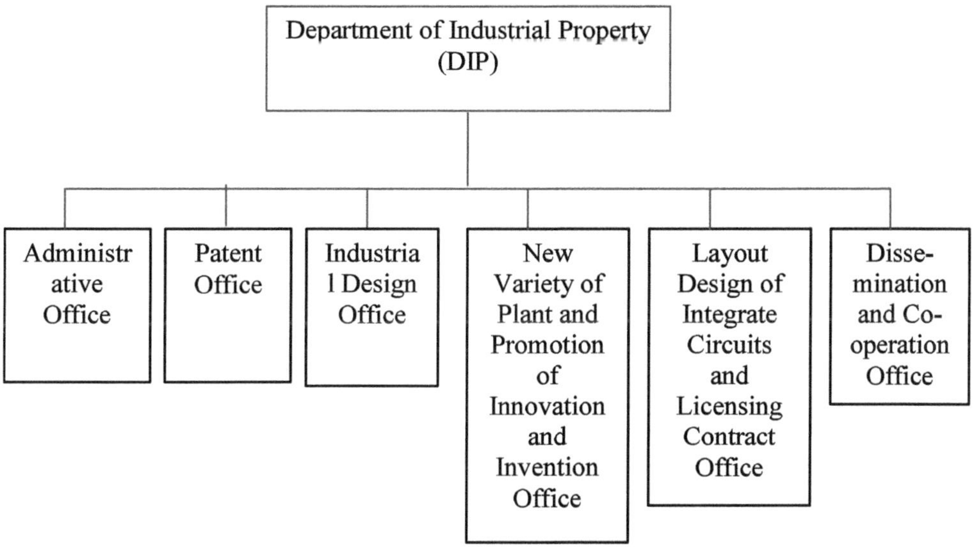

Figure 17.1: Organization chart of Department of Industrial Property (DIP)

Beside this, we has IP committee current the Ministry of Commerce standing as a Chairman, mention the Sub-decree on the National Committee Governing Intellectual Property (14 components) of the Kingdom of Cambodia and it was adopted on 18 September, 2008. The members are as under:

- ☆ Ministry of Commerce (Chief)
- ☆ Ministry of Industry Mines and Energy (Deputy Chief)
- ☆ Ministry of Culture and Fine Art (Deputy Chief)
- ☆ Representative of Council of Ministers (Member)
- ☆ Representative of Ministry of Economy and Finance (Member)
- ☆ Representative of Ministry of Information (Member)
- ☆ Representative of Post and Telecommunication (Member)
- ☆ Representative of Health (Member)
- ☆ Representative of Ministry of Agriculture, Forestry and Fisheries (Member)
- ☆ Representative of Ministry of Environment (Member)
- ☆ Representative of Ministry of Justice (Member)
- ☆ Representative of Ministry of Education Youth and Sport (Member)

- Representative of Ministry of Interior (Member)
- Representative of Tourist (Member)

Implementation of the Law on Patents, Utility Model Certificates and Industrial Designs

In order to have an effect on this policy the law on Patent, Utility Model Certificates and Industrial Design was established in 2003.

Presently, we had the Law on Patent, Utility Model Certificate and Industrial Designs and our Office just open the door for receiving applications for inventions as well as Industrial Designs.

Department of Industrial Property (DIP) has operated the process of grant of Patent, Utility Model Certificates and Registration of Industrial Designs since 29 June, 2006.

Statistics of December 2010

1. Patent

   Received 106 Applications, all from Overseas.
2. Industrial Design

   Received 133 Applications from Local 23 and Overseas 110.

   Registration 105 Applications for Local 19 and for Overseas 86.

The Problems with IP in Cambodia

The Kingdom of Cambodia is facing many problems of enforcement the IP. People have little knowledge about IP and there is lack of public awareness on intellectual property.

Our Department of Industry Property is facing many difficulties such as lack of technical aspect and materials, professional officials do not have enough experiences and human resources development is also needed for upgrading the official's knowledge and skill in various sections of the process of receiving application, formality check, search and examination, grant of patent, registration of Industrial designs and administration of patent and industrial design as well.

Future Objective

On other hand, we are upgrading the knowledge and skills for professional officials in the field of Intellectual Property by sending them to attend the training courses, seminars and workshop in this field, locally or abroad, strengthening and widening the cooperation with other International Organizations and other Intellectual Property Offices in exchanging information and experiences for the development of Industrial Property System.

Chapter 18

Lesser Known Horticultural Products as Geographical Indications in India

S.K. Soam and Mujarrab Hussain
National Academy of Agricultural Research Management,
Hyderabad – 500 407, India
E-mail: soamjee@gmail.com

ABSTRACT

Products registered as geographical indications are beyond doubt function as powerful economic tools, thereby influence livelihood situations leading to transformed socio-economic condition. This may happen to significant number of rural population in India also. But unfortunately the trend of registration of horticultural products in the country has been far less than the satisfactory. The most important reason for this is lack of awareness and secondly failure of proactive actions from both private and public institutions; in case of unorganized sector the responsibility mainly lies with public organizations, but except a few cases these have not been very responsive. Public system should immediately take lead in facilitating the registration of well-known products, while ground should be prepared for lesser-known products. These actions are to be coupled with other policy interventions also. The present paper record several less-known but locally very popular products from different regions of India, which can be powerful economic tool as GI; for accrual of maximum advantages, several policy implications have also been identified and appropriate recommendations have also been made.

***Keywords**: Geographical indication, Biodiversity, Socio-economic upliftment, IPR.*

Introduction

In India, the process of protection of geographical indications was initiated with the inception of the legislative Act called "The Geographical Indications of Goods (Registration and Protection) Act 1999", that came into force since September 2003, till September 2010, 22 horticultural products have been registered, which is 14 per cent of total registrations, 6 are under consideration.. Post-legislative responsibilities include identification of potential products, conducting socio-economic studies, initiating registration, and involving scientific and technical interventions for market success. The present research work identifies lesser-known horticultural goods as Geographical Indications (GI) in several states in the country, and also addresses some of the issues pertaining to use of GI as a tool for socio-economic development of the resource poor rural people.

The main objective is to share information for creating awareness among the local key stakeholders, and dispersing the knowledge about the product to the probable domestic and international consumers is another purpose of this paper. Third objective is to draw attention of corporate sector for association with the trade worthy products. Fourth objective is to draw attention towards scientific and technical necessity for establishing geographical link, uniqueness, quality up gradation and development of inspection measures. For the products of lesser-known nature, the registration as GI requires special efforts such as mobilizing people, arrangement of financial, technical and legal support.

Methodology of Data Collection

The collection of data and related primary information about the potential GI is still a challenge because of persisting ignorance about this new form of community intellectual property. Therefore in the present project, a novel method is used for collection of data and primary information. A two-hour session was introduced on GI in various training programs conducted by organizations based at Hyderabad and elsewhere in the country. After making the trainees fully aware about GI, they were given a format to provide information about the products and their characteristics that can serve as GI. Another platform for data collection was United Nations Conference on Trade and Development (UNCTAD) funded project, where small brainstorming sessions were conducted through out the country. The data collection activity was mainly done during 2006-2009 involving about 700 persons from 25 states in the country.

Results

The data obtained from the respondents was analyzed and clustered with respect to various classifying parameters. Therefore the results have been presented as per the data grouping, it include the broad classification of state/zones in the country, nature of the product with brief description and mainly under GI class 30 (spices and condiments) and 31 (Fruit and vegetables) of schedule IV of the Indian GI Act, some products under class. 29 *i.e.* jelly and other preserved fruits and class 32 *i.e.* fruit drinks have also been discussed.

The results are presented for various states in following order- Andhra Pradesh (AP), Karnataka, Tamil Nadu, Kerala Madhya Pradesh (MP), Maharashtra, Rajasthan, north eastern region consisting of West Bengal (WB), Bihar, Assam, Sikkim and Arunachal Pradesh; northern and hilly region consisting of Uttar Pradesh (U.P.), Uttarakhand, Himachal Pradesh (H.P.), Jammu and Kashmir (J&K) and Punjab.

Table 18.1: Andhra Pradesh: Fruits and Other Allied Products

Name	*Description*	*Place*
Amrita chakarkelly	Local banana	Khamam
Balanagar sithaphal	Custard apple local variety	Mahboobnagar, Medak, Nalgonda, Warangal
Kohir jam	Sweet and fleshy guava	Medak
Cherukurasam	Juicy mango variety	Krishna
Chenna and Pedda rasalloo	Reputed mango variety	Andhra region
Kallapur mango	Local mango variety	Mahaboobnagar
Swarnrekha mango	Juicy mangoes	Srikakulum, Vizinagram and Vizag
Nujiveedu mango	Juicy and tasty mangoes	Krishna
Allumpur benishan	Quality mango	Mahaboobnagar
Veerabalu/Veerapalli	Benisan mango variety with yellow red tinge and specific taste	Kadapa, Chitoor
Ulavapadu mango	Good keeping quality and taste	Prakasham
Damalacherulu khader	Good taste mangoes	Chitoor
Chintalapudi punasa	Local variety of mango	Krishna
Banganpally (Benishan)	Reputed, juicy and tasty mango variety with high keeping quality	Banganpally village, Kurnool district but grown in most AP
Palakollur battai	Local sweet orange variety	West Godavari
Pulivendla mosambi	High quality sweet oranges	Kadapa
Pala sapota	Very sweet and high fiber content	Kurnool
Alamanda mango jelly	Mango jelly	Alananda and Bheemali of Vizianagram
Allam murabba	Preserved medicinal ginger	Hyderabad
Vetapalem cashew	Local cashew nut	Prakasham
Palasa cashew	Big size of cashew apple	Srikakulam
Chennur pan	Quality betel vine leaves	Kadapa

Table 18.2: Andhra Pradesh: Vegetables

Name	*Description*	*Place*
Krishnapuram Vulli or KP Onion	Pungent, medium size, medicinal value and best for sambhar	Kadapa but known to be grown in Tamil Nadu also
Guntur gongura	Quality mesta used in pickle and sambhar	Guntur
Madanapalli tomatoes	Availability throughout the year	Chittoor
Sareekanda	Rrhizomatous vegetable	Srikakulum
Kakarakaya	Bitter gourd of high bitterness found in black soil	Kurnool, Anantpur
Jagatial long	Local ridge gourd	Karimnagar
Bara batlu	Greens for curry preparation	Visakhapatnam
Tanuku venkaya	Green brinjal	West Godavari
Indukurpeta venkaya	Brinjal,soft, smooth and tasty	Indukurpeta village of East Godavari
Pounada venkaya	Brinjal with taste like cheese	Srikakulam
Baptala brinjal	Round light purple color brinjal	Guntur
Mula venkaya	Local brinjal variety	Medak
Bulli venkaya	Local brinjal variety	Chittoor
Poluru brinjal	Spiny green fruits with white strips	Kurnool

Table 18.3: Andhra Pradesh: Spices and Condiments

Name	*Description*	*Place*
Mydukur pasupu and Tekurpet turmeric	Long type turmeric, bright in color, high keeping quality	Kadapa
Duggirala turmeric	Local popular variety	Krishna and Guntur
Armoor turmeric	Local popular variety	Nizamabad
Gokavaram chintapandu	Less sour tamarind	East Godavari
Rampachoda	Tamarind grown in tribal region	East Godavari
Charu mirchi	Chili variety bearing small fruits with high pungency	Srikakulum
Warangal chapata	Traditional chili variety of deep color useful for pickle making	Warrangal
Guntur karam	Pungent, dark red colored chilies	Guntur
Gollapalli mirchi	Local variety of chili	East Godavari
Kohir zinger	Red color zinger	Medak
Siddipet allam	Zinger variety	Siddipet, Medak

Table 18.4: Karnataka: Fruits and Vegetables

Name	*Description*	*Place*
Nanjanagud rasabale	Banana with unique taste	Nanjanagud, Mysore
Munnavalli banana	Local banana genotype	Belgaum
Coorg orange	Specific orange	Coorg region
Matigulla brinjal	Green round thorny brinjal variety	Dakshin Kannada and Udupi districts
Porrigulla brinjal	Round brinjal	Udupi
Sakrepatha jackfruit	Good taste, color	Chikmagalur

Table 18.5: Karnataka: Coffee, Spices and Flowers

Name	*Description*	*Place*
Giri coffee	Arabica coffee with typical aroma	Chikmagalur
Chikmagalur coffee	High crown coffee	Chikmagalur
Coorg coffee	Reputed coffee	Coorg region
Bababudangiri coffee	Good quality coffee	Chikmagalur
Byadgi chilli	Dry chili with extreme pungency	Haveri, Gadag and Dharwad
Sankeshwar chilli	Local chili variety	Belgaum
Ajjampura coriander	Golden color excellent quality	Chikmagalur
Pedavagi chillies	Specific variety of chilies	Bijapur
Mysore mallige	Highly fragrant jasmine with bold and round flowers	Mysore
Udupi mallige	Fragrant jasmine variety	Udupi

Table 18.6: Tamil Nadu: Miscellaneous Horticultural Products

Name	*Description*	*Place*
Nilgiri tea	Good quality tea	Nilgiri ranges
Erode turmeric	Specific kind of turmeric	Erode
Madurai malli	Jasmine flower	Madurai
Salem mulgovo	Specific mangoes	Salem, Dharmapuri
Vayal vazhi	Banana variety with thick skin	Trichy
Virupakshi banana	Good taste banana, resistant to certain diseases	Palani hills
Kumbhkonam arthur	Betel vine famous like banarasi pan	Kumbhkonam

Table 18.7: Kerala: Fruits and Vegetables

Name	*Description*	*Place*
Chavakkad orange	Dwarf coconut cultivar with orange nut, good for nut water	Chavakkad of Trichur district
Chemparathi varikka	Indigenous variety of jackfruit with red perianth	Kollam
Kurukkan	Salt tolerant mango, polyembryonic with dwarfing effect	Kannur, Kasargod, Kozikode

Table 18.8: Kerala: Spices, Condiments and Tea

Name	*Description*	*Place*
Karuvilanchy black pepper	Bold seeds, high oil content, resistant to various stresses	Kollam
Telichery black pepper	Dried black pepper	Telichery, Kannur
Alleppey turmeric	High curcumin content	Wayanad
Wayanad turmeric	Known for specific characteristics	Wayanad
Kuruppumpadi ginger	Highly fibrous, diseases resistance	Ernakulum
Alleppy ginger	Known for specific characteristics	Alleppy
Alleppy green cardamom	Quality cardamom	Alleppy
Wayanadan tea	Tea leaves	Wayanad

Table 18.9: Madhya Pradesh: Spices and Condiments

The important spices are 'Kumbhraj dhania', coriander of good flavour and good oil content found in Guna and Rajgarh; a good quality fenugreek at Jaora, Ratlam and Neemach, and Amleta and Mahadev are good quality local garlic from Mandsaur and Neemach.

Table 18.10: Madhya Pradesh: Fruits and Vegetables

Name	*Description*	*Place*
Chappara seetaphal	Large sized custard apple with long shelf life	Seoni
Kathiawadi mango	Big size mango	Alirajpur in Jhabua district
Sundaraja mango	Sucking type mango	Rewa
Malwa potato	Low cholesterol potato	Malwa region
Baigani Arhar	White color bold seeded red gram pods used as vegetable	Mandla
Selana ki kakdi	Good quality cucumber	Ratlam

Table 18.11: Maharashtra: Fruits

Name	*Description*	*Place*
Bahadoli jamun	Big size Indian blackberry	Thane
Bhagava anar	Pomegranate of typical color	Malegaon
Nasik grapes	Small sweet grapes	Nasik, Pune, Sangali, Satara, Ahmednagar
Katol sweet orange	Sweet orange fruit	Vidarbha region
Ahmednagar sweet orange	Sweet orange fruit	Ahmednagar
Hapus mango	Mango variety like alphonso	Konkan
Keshar mango	Mango variety of specific taste and aroma	Beed, Latur, Marathwada region
Vasai banana	Small size banana	Vasai in Thane district
Basrai banana	Dwarf banana	Jalgaon
Jalgaon banana	Banana with typical taste, flavour	Jalgaon
Sukeli	Ripe big banana with specific taste	Vasai in Thane district
Bhusawal kela	Fruit banana	Bhusawal
Ardhapuri banana	Sweet, specific aroma, yellow color fruits of big size	Ardhapur in Nanded district

Rajasthan

Rajasthan is bestowed with several kinds of products e.g. 'mehandi of Sojat', which is henna for body painting, grown and prepared at Saujat town in Pali district. Coriander of Kota district, nagauri methi, fenugreek from Nagaur, bhichiwara ginger from Bhichiwara in Dungarpur district. Phalodi chilies from 'Phalodi region' in Jodhpur are famous due to specific pungency. Dungarpur zinger, a local zinger variety from Dungarpur and methania chilies, which are long chilies are used to give color to dishes in Jalore and Jodhpur districts.

Several fruits and flowers are also famous from Rajasthan such as rose flowers from Pushkar are good for gulkand making, 'Dhomas rose' of Haldighati, Udaipur are reputed for oil. Jhalawar orange from Jhalawar; banswara mangoes from Banswara are known for specific sour and sweet taste mango, *Aam papad* made of these mangoes is popular and can be protected under class 29. Chomu ber (Zizyphous) from Chomu tehsil, Jaipur is known for fruit borer resistance.

Discussion

Several papers are available citing number of benefits of GI registration of a product available in Europe (Rangnekar, 2004), Asia-pacific (Swarnim Waglé,2007) and India (Nagarajan, 2007). Community ownership is major objective of GI protection e.g. sea-buckthorn (*Hippophae rhamnoides*) is a nitrogen fixing non-legume shrub found in Leh-Ladakh and locally known as 'Tserma' or 'Jadoo paudha', the indigenously produced juice from its berries have been commercially exploited under a probable

Table 18.12: Horticultural Products in North-eastern States

State	*Product*	*Description*	*Place*
WB	Baraipur guava	Guava variety	Baraipur block
WB	Malda or Fazali	Locally known mangoes	Malda
WB	Laxman bhog	Locally known mangoes	Malda
WB	Watermelon	Blackish watermelon	Sagar island
WB	Martamam	Unique kind of banana	Hoogly and Nadia
WB	Darjeeling orange	Small size very sweet	Darjeeling
Bihar	Makhana	Specific dry nut grown in ponds and used after roasting	Bhagalpur
Bihar	Muzaffarpur litchi	Famous litchi fruit	Muzaffarpur,
Manipur	Lakadong turmeric	Locally known turmeric	Meghalaya also
Assam and Arunachal	Malbhog banana	Banana variety	Kamrup, Darrang, Nalbari, Nowgong
Assam	Sanchi pan	Variety of betel leaf	Nogaon, Morigaon
Assam	Pine apple	Specific flavour	Guwahati
Sikkim	Temi tea	Locally known good tea	Temi, north Sikkim

Table 18.13: Northern and Hilly States

State	*Product*	*Description*	*Place*
U.P.	Langda	Big size mango variety	Eastern UP
U.P.	Chaunsa	Juicy mango variety	Eastern UP
U.P.	Gulawthi ke ber	Zizyphous fruit	Gulawthi, GZB distt
U.P.	Allahabad safeda	Guava variety of repute	Allahabad
U.P.	Allahabad surkha	Pink flashed guava	Allahabad
U.P.	Pratapgarh aonla	Quality Indian gooseberry	Pratapgarh
U.P.	Jaunpuri muli	Large sized tasty radish	Jaunpur
U.P.	Mahoba paan	Betel leaf	Mahoba
Punjab	Kinnow	Orange kind of fruit	West Punjab
Uttarakhand	Beninag tea	Tea with specific taste	Almora
Uttarakhand	Harshil apples	Locally famous apple	Uttarkashi
Uttarakhand	Ramnagar litchi	Locally popular litchi fruit	Nainital
H.P.	Kangra tea	Tea of repute	Kangra
H.P.	Kinnauri apple	Dark red late harvested	Kinnaur
H.P.	Pahari aloo	Less sweet potatoes	All HP in summer
J&K	Kashmiri Kesar	Saffron with special flavour from Doda distt	Phulwara, and Kaishtwari tehsil
J&K	Tsetalulu	A locally made fruit drink	Leh, Ladakh

GI name 'Leh Berry' with no community involvement. For the lesser known horticultural products, very specific advantages that can be accrued to the society through GI registration are:

- ✰ Organizing the producers as Commodity Interest Group (CIG)
- ✰ Sustainable conservation of agro-biodiversity through bringing the products into main supply chain
- ✰ Popularizing the restricted rural brands among new areas and enhancing its consumption sphere
- ✰ Inculcating the entrepreneurship attitude within rural resource poor people
- ✰ Marketing the rural specialty through corporate association such as TATA nilgiri tea or Amul Alphonso ice cream
- ✰ Commercialization of new primary and secondary products and to make them normally traded commodities e.g. constant availability of Lakadong haldi or Nagauri methi at major spices markets.

Most of the products are known in production area only, a few at one or two districts in the state and very few such as Banganpally mango are known through out state only. Soam, 2005 documented several products as potential GIs, few lesser known of them such as Makhana, Kangra tea, Pokkali, Navara, Jeerakshala and Gandhakshala rice have been registered in later years. Out of the discussed products in this paper, 14 have been registered as given in the Table 18.14, Bydagi chilli is under examination phase.

Table 18.14

Kangra Tea	Mysore Malligae	Nilgiri Tea
Coorg Orange	Udupi Malligae	Virupakshi Hill Banana
Nanjanagud Banana	Alleppey Green Cardamom	Bydagi Chilli
Allahabad Surkha	Laxman Bhog Mango	Guntur Sannam Chilli
Tellicherry Pepper	Fazli Mango	Nashik Grapes

Therefore discussions on little known products are of utmost importance. Many products as Temi tea or Beninag tea are as good as any other popular tea; these products have direct commercial value; only challenge is that they may remain lesser known even after registration as happened with Kangra tea because of competition with other valued and registered products in the same sector i.e. Darjeeling tea and Assam tea. In some cases alternative products as mango pulp or jelly, juices, nursery plants or propagating material may have good commercial potential. But several others as local brinjal verities from AP and Karnataka may not have direct commercial value but these are important Plant Genetic Resources (PGRs). The PGRs can be used for varietals improvements and in consultation with breeders these can be given affix or suffix to GI name and later can be registered as verities under plant variety protection provisions, therefore a win-win situation for both breeders and owners of GI. The Biodiversity Management Committees (BMC) set up under biological diversity Act,

2002 must take the lead in GI portfolio development of such products. For example, KP onion from AP could be a good case for variety improvement or alteration. Due to its typical pungency it is reputed for use in specific south Indian recipe such as 'sambhar'. Latest research reports that Quantitative Trait Loci (QTL) controls pungency; there is a significant associations of both pungency and bulb soluble solids content, and genetic markers may provide a practical means to select for lower pungency without correlated selection for lowered solids (John McCallum et al, 2007), thereby keep other basic characters intact.

Establishing uniqueness among lesser known products is another challenging task e.g. mydukur pasupu or turmeric and tekumpet turmeric both are found in Kadapa/Chitoor districts of AP but scientific intervention is necessary to check whether both are same or different. Similar to this, recently Thomas et al (2006) through biochemical characterization established that Indian cardamom is different and better than the cardamom from Sri Lanka and Guatemala. Scientific and technical interventions are also required to ensure product traceability in supply chain, product characteristics traceability through technological means. Good example for this is 'Jamaica' where; using GIS and GPS applications, traceability for hot pepper has been done (Anonymous, 2006).

Conclusion

Antiquity can certainly not be essential requirement in horticultural goods; some products have been evolved in so many years and are famous since ages. The main purpose of GI registry is commercial exploitation for economic upliftment of producers' community; in achieving this goal the other socio-economic, ecological and environmental aspects must not be overlooked. Some of these aspects are conservation of fragile species and protection of rural resource poor producers. Government may take appropriate measures, lest in the wake of economic exploitation, the original producers are replaced by the large corporate sector. The rural poor may also face in-house competition with big players in the same geographical region in the form of large-scale specialty production, corporate farming or contract farming. The scientific and technical institutions have challenges for not only improving the GI goods but also developing scientific inspection mechanism. Scientific deliveries as a means of protecting and promoting community intellectual property rights are other bigger challenges for scientific institutions that can support other institutions.

Acknowledgements

The first author is grateful to Director, NAARM, Hyderabad, for receiving all necessary support from the institute, and to United Nations Conference on Trade and Development (UNCTAD) for providing the consultancy project for the work.

References

Anonymous, 2006. ICT update a current awareness. Bulletin for ACP Agriculture, issue 32, June 2006, CTA, Wageningen.

John McCallum, Meeghan Pither-Joyce, Martin Shaw, Fernand Kenel, Sheree Davis, Ruth Butler, John Scheffer, Jernej Jakse and Michael J. Havey, 2007. Genetic mapping of sulfur assimilation genes reveals a QTL for onion bulb pungency. Theoretical and Applied Genetics, 114 (5), pp 815-822.

Nagarajan, S., 2007. Geographical indications and agriculture-related intellectual property rights issues. Current Science, 92(2), pp167-171.

Rangnekar Dwijen, 2004. The Socio-economics of Geographical Indications- A Review of Empirical Evidences from Europe. United Nations Conference on Trade and Development, Geneva, May 2004, issue paper no. 8, 52p.

Soam, S.K., 2005. Analysis of prospective geographical indications of India. Journal of World Intellectual Property, 8(5), pp 679-704.

Swarnim Waglé, 2007. Geographical indications as trade-related intellectual property: relevance and implications for human development in Asia-Pacific. Discussion paper, Asia-Pacific Trade and Investment Initiative, January 2007, UNDP Regional Centre, Colombo, 42p.

Thomas, Kizhakkayil, Zachariah, Syamkumar and Sasikumar, 2006. Comparative quality characterization and molecular profiling of Indian, Sri Lankan and Guatemalan cardamoms. Food, Agriculture and Environment, 4 (2), pp 59-63

Chapter 19
Ecosystem for IPR in India Especially for Patents

Yashawant Dev Panwar
Scientist-D,
Patent Facilitating Centre (PFC),
Technology Information Forecasting and Assessment Council (TIFAC),
(Department of Science and Technology, Government of India),
A Wing Vishwakarma Bhavan,
Shaheed Jeet Singh Marg, New Delhi – 110 016, India
E-mail: yashawant@tifac.org.in

ABSTRACT

Intellectual Property Rights (IPR) system in India is not new, but only after 1995 when India became part of World Trade Organization (WTO), Intellectual Property (IP) activities in India have assumed greater importance and widespread role at all levels. Basically, the ecosystem of IP consists of IPR laws, their administering authorities, IP facilitation mechanism, awareness, a part of academic schedule in many forms like curriculum, research, professional courses, papers etc., legal system including IPR agents and attorneys, judges, courts, litigators and IP creators including scientists, technologists, authors, musicians, keepers of traditional art, industry, services or trade. Laws enacted with respect to IPR in the last 15 years include: New Trademarks law, New Design Law, Plant Variety protection, IC Layout Design, Geographical Indications. Beside these laws there have been amendments in Patent Act and Copyright Act. Apart from these laws an Intellectual Property Appellate Board (IPAB) has also been created to resolve disputes prior to going to court. IP offices have been modernized. Other than enactment and enforcement of various laws in the country after joining WTO and TRIPS, India has also acceded to various international treaties like Paris Convention, Patent Cooperation Treaty (PCT), Budapest Treaty, Universal Copyright Convention, Phonogram Convention and is planning to accede to Madrid Protocol too. The significant resources have been employed by various agencies to create appropriate environment for protection and exploitation of IP in the country and this has been discussed in this chapter in detail, particularly focusing on issues relating to patents in general and patenting facilities.

***Keywords**: India, IPR laws, IP ecosystem, IP policies, Patent granting system, Patent facilitation.*

Introduction

In this paper, a holistic approach has been adopted to elaborate the Intellectual Property (IP) ecosystem in India. It is possible for any country to draw adequate advantage of the system if the same is supported by education, necessary creations (innovation), laws, raw material (in this case existing knowledge and existence of physical material), finished goods (inventions, designs, literary and all other kind of creative work), protection preservation and exploitation (protecting IP through patent and other IP forms and maintain the same). Without enforcement and exploitation, the value of IP would not be worth talking. On emergence of WTO in 1995 and India being one of the members, there have been lot of challenges and opportunities for all. General feeling at that time was that there are more challenges then the opportunities. As time passed, all the challenges are being successfully taken care of baring few, but opportunities are not being fully exploited An half-hearted efforts are reflected in all statistics pertaining to Intellectual Property Rights (IPR).

The legal ecosystem in India has been recreated by enacting various new laws like Geographical Indications, New Plant Variety Protection, the Semiconductor Lay out Design Act; and amendments of Trademarks, Registration of Industrial Design, The Patent Act and The Copyright Act. Besides these, there are other laws which influence the sphere of IP. These are The Seed Act, National Biodiversity Act, The Competition Act, Information Technology Act etc. Except the Semiconductor and IC layout design act, all the laws are operational through newly set-up authorities/ offices or by carrying out modernisation of existing offices/authorities. Courts have been delivering justice as fast as possible in the matters of IP and it's a great boost to the whole Indian IP legal system. However, still lots of ground is to be covered towards efficient enforcement of the system relating to IPR.

Accession to various international treaties has enabled Indians to obtain IP protection in other countries with the privileges offered by them. At the same time citizens from member countries are also availing such facilities and getting protection to their intellectual prowess in India. Government has come up with various schemes to foster innovation and creativity in the country. Some of the schemes like INSPIRE of DST, TePP of DSIR, TreMAP of TIFAC, NMCP by ministry of MSME and various such schemes by other ministries would definitely strengthen ecosystem of innovation at large.

Lack of manpower in the area of IPR was a great dampener at one point of time for the country. Sincere efforts in this direction have been able to provide manpower in the field of IPR through a number of schemes like WOS-C implemented by PFC-TIFAC, courses run by National Law Schools and other law schools, NIIPM under CGPTDM, IGNOU and other private efforts like GIPP etc. There has been considerable number of IP practitioners trained on Job internship by IP Cells of government agencies as well as private industries apart from formal courses. However, as yet skill levels need to be handled vigorously by providing right kind of training aggressively. Facilitating mechanism and awareness creation has its role in development of IP culture in the country. Agencies like PFC spearheaded the movement of IP/Patent awareness in the country. PFC has been instrumental in devising new models of handling IP in the countries in the area of awareness, training, and facilitation of filing IP in India and abroad. Best of these models have been adopted by other agencies at large. Other agencies like NRDC, MHRD, Ministry of MSME has also put up considerable effort for creation of awareness. Even patent office, other ministries has also played their part of role in awareness. Institution like DRDO, ISRO, CSIR, DAE have their mechanisms of advance training of IPR to their scientists and technologists along with full support system for protection of IP emanating from institutions belongings to their set-up.

Indian will have to wait to see sizable economic impact specially the benefits of IPR system. Raising bar to the next level of IP management and exploitation is the need of the hour. Ecosystem needs orientation towards responding to the emergent techno legal environment in the country. In this paper, attempts have been made to put all efforts together to provide input to the thinking process at all levels as well as the single window info on IP.

IP Granting System

Other than copyrights, trade secrets and to some extent trademarks, all other IPRs namely patents, industrial design, IC layout design, Geographical Indications, new plant varieties, application, grant and registration are a must to acquire the rights. To have better understanding of IPR, Table 19.1 explains the type of works to get protection under different forms of IP in India.

Table 19.1

Sl.No.	*Form of IP*	*Type of Work Protectable*	*Term of Protection*
1.	Patents	Inventions (new product or process)	20 years from date of filing
2.	Copyrights	☆ Literary, dramatic and musical work ☆ Artistic work ☆ Cinematographic films which includes sound track and video films ☆ Record-any disc, tape, perforated roll or other device	Authors life plus 60 years for litrary work and 60 years from publication of work in all cases 25 years from year of broad cast for broadcasting rights
3.	Industrial Design	Feature of shape, configuration, pattern, ornament or composition of lines or colors applied to an article in 2D and 3D both. In other words appearance.	15 Years
4.	Trademarks	Distinct sign; combination of words, letters, numerals, consist of drawings, symbols, 3D colors, combination of colors. Can also be registered as service marks, collective marks and certification marks.	10 years and can be renewed for indefinite period in the cycle of 10 years

Contd...

Table 19.1–Contd...

Sl.No.	*Form of IP*	*Type of Work Protectable*	*Term of Protection*
5.	Geographical Indications	Names associated with goods (Agriculture, natural or manufactured goods) wherein given quality reputation or other characteristics of such good is essentially attributable to its geographical origin.	10 years and can be renewed for indefinite period in the cycle of 10 years
6.	Semiconductor IC Layout Design	Layout design of transistors and other circuitry elements including lead wires forming on a semiconductor or an insulating material or inside the semiconductor material and designed to perform an electronic circuitry.	10 years from filing or date of first commercial exploitation whichever is earlier.
7.	Plant Variety Protection	Breeders Varieties (New) Extant varieties including Farmers Varieties	**Trees and vines 18 years** from the date of registration. All other cases **15 years from** date of registration
8.	Trade secret	Every information which is not disclosed	No time limit as long as it is kept secret

Presently following laws are in place for IPR in India and offices responsible for grant of rights.

IP offices under the Office of Controller General of Patents, Designs and Trademarks with four branches are at Delhi, Mumbai, Kolkata and Chennai. In addition to the four places, trademarks registry is in Ahmadabad also. GI registry is in Chennai only. These offices are under the Department of Industrial Policy and Promotion (DIPP), Ministry of Industry and Commerce. Patent office administers the patent. The Patent Act, design registration is handled by Kolkata patent office but application can be filed at any of the four patent offices. Trademarks application is to be filed in Trademark Registry which are collocated with all four patent offices and also in Ahmadabad. The GI Registry is co-located with patent office at Chennai.

These four forms of IP are administered under their respective Acts as follows:

- ☆ **Patents :** The Patents Act, 1970. It was amended in March, 1999, 2002 and 2005.
- ☆ **Design** : A Design Act, 2000.
- ☆ **Trade Marks :** A new Trademarks Act, 1999.
- ☆ **Geographical indications (GI) :** The Geographical Indication of Goods (Registration and Protection) Act, 1999.

Copyright office is located in New Delhi and is under the Ministry of Human Resources and Development (MHRD) and is administered under The Copyright Act, 1957 as amended in 1983,1984, 1992,1994 and 1999 and the Copyright Rules, 1958.

IC layout design registry has been set-up under Department of Information Technology, Ministry of Information Technology (MIT). The law for the same is The Semiconductor Integrated Circuit Layout Design Act, 2000.

The New Plant Variety and Farmers Rights Authority registers new plant varieties. This is under Department of Agriculture Research, Ministry of Agriculture and is located in New Delhi. The law for the same is The New Plant Variety Protection and Farmers Rights Act 2001

Protection of undisclosed information: No exclusive legislation exists but the matter would be generally covered under the Contract law (Contract Act 1872).

Conditions to be Fulfilled for Filing Patent for Grant

Prerequisites for Invention to Qualify for Patent

Patents are granted for a new process or product irrespective of area of technology. The patent act defines invention which could be patented "Invention means a **new product or process** involving an **inventive step** and capable of **Industrial Application**. Newness is assessed in global context, it should not have been published or used anywhere in the world prior to filing of patent application. The second term inventiveness is defined as "invention that involves technical advance as compared to existing knowledge or having economic significance or both and that makes the invention not obvious to a person skilled in the art". The third term or condition is that it is capable of industrial application, which means it can be used or made in Industry.

The patent act of India prohibits patenting of inventions which are defined as inventions not patentable. Those are as follows:

1. An invention which is frivolous or which claims anything obviously contrary to well established natural laws;
2. An invention the primary or intended use or commercial exploitation of which could be contrary to public order or morality or which causes serious prejudice to human, animal or plant life or health or to the environment;
3. The mere discovery of scientific principle or the formulation of an abstract theory or discovery of any living thing or non-living substance occurring in nature;
4. The mere discovery of a new form of a known substance which does not result in enhancement of the known efficacy of that substance or the mere discovery of any new property or new use for a known substance or of the mere use of a known process, machine or apparatus unless such known process results in a new product or employs at least one new reactant;

   Explanation: For the purpose of this clause, salts, esters, ethers, polymorphs, metabolites, pure form, particle size, isomers, mixtures of isomers, complexes, combinations and other derivatives of known substance shall be considered to be the same substance, unless they differ significantly in properties with regards to efficacy.
5. A substance obtained by mere admixture resulting only in the aggregation of the properties of the components thereof or a process for producing such substance;

7. The mere arrangement or re-arrangement or duplication of known devices each functioning independently of one another in a known way;
8. A method of agriculture or horticulture;
9. Any process for medicinal, surgical, curative, prophylactic, diagnostic, therapeutic or other treatment of human beings or any process for a similar treatment of animals to render them free of disease or to increase their economic value or that of their products;
10. Plants and animals in whole or any part thereof other than microorganisms but including seeds, varieties and species and essentially biological processes for production or propagation of plants and animals;
11. A mathematical or business method or a computer program per se or algorithms;
12. A literary, dramatic, musical or artistic work or any other aesthetic creation whatsoever including cinematographic works and television productions;
13. A mere scheme or rule or method of performing mental act or method of playing game;
14. A presentation of information
15. Topography of integrated circuits
16. An invention which, in effect, is traditional knowledge or which is an aggregation or duplication of known properties of traditionally known component or components
17. Inventions relating to atomic energy

Other than Above Provisions; when Patents Relating to Biological Materials/ Microorganisms are Filed following compliance has to be Adhered

If patent application involves microorganism, these microorganism to be deposited in International Depository Authority (IDA) notified under Budapest Treaty. In India IMTECH Chandigarh has been recognised as IDA, under Budapest Treaty. The microorganism to be deposited before the filing of patent application and access to public should only be available after filing of patent. The IDA would assign an accession number on deposition of sample. This accession number is to be quoted in patent application.

In addition to accession number it is mandatory to mention the geographical origin of the microorganism. Failing which or wrong disclosure of origin, patent can be cancelled.

It also mandatory to obtain the permission from National Biodiversity Authority (NBA). As per section 6 of biodiversity act; permission to be obtain from NBA before the sealing of the patent, means before grant of patent. This provision is not restricted to microorganism only but applies to all biological materials referred in patent documents and some claims have been made on them or process involving them.

All kind of microorganisms are not accepted by IMTECH, in such cases IDA can be identified from list of 40 odd IDA across the globe which are recognised under

Budapest Treaty. Selection can be made based upon facilities available and cost charges by the IDAs.

Process of Filing and Obtaining Patents

The filing of patent is possible in India in hard copy at patent office or e-filing through portal of Patent Office www.ipindia.nic.in. Patent are filed with patent office with provisional or complete specification. Patent office publishes patent after 18 months of filing patent application. If applicant requires early publication the request has to be made by application to patent office with the payment of prescribed fee to the patent office. It is also mandatory to file the request for examination of patent. If no request for examination is filed, patent will not be examined. This request has to be filed before the expiry of 48 months from date of filing. The application will be put in queue after the publication of patent application or filing of request for examination whichever is later. On getting its turn in queue, Patent office will issue first examination report(FER), this include both formal and technical examination. Under technical examination they will assess the patent claims against all prerequisites for invention to qualify for patent. In turn applicant will respond to the queries of patent office and this application has to put in order within 12 months of issuance of FER. Means has been rejected or granted within the 12 months from issuance of FER, if required applicant or their attorney may be given opportunity to defend the claims through hearing. After the grant, patent has to be maintained by paying renewal fee.

Indian system provides pre grant and post grant opposition system. Any party can oppose the patent after the publication and before the grant by writing to controller of patents. But after the grant of patent it can be opposed by interested party within one year of grant. Decision of the Controller of patent could be challenged before the Appellate Board. If not satisfied, parties can go to courts.

Filing of Patents in Other Countries

It would be mandatory for Indians to file patent first in India and wait for 6 months before one files patent in other country or else obtain permission from patent office to file patent application in other country. There would be three routs of filing patent application in other country.

Paris Convention

If applicant file patent in any member country of Paris Convention within 12 months of filing of patent application in India the priority date in those country will be filing of patent in India.

Patent Cooperation Treaty (PCT)

If applicant adopt this rout one can file PCT International application within 12 months of filing of patent in India. This is an filing mechanism only, for the grant of patent one will have to file national phase applications in the countries of interest in most cases within 30 to 31 months from date of priority.

One can file in other country without claiming the priority before the publication of patent application by patent office in India.

Modernization of Indian Patent Office

Significant work has been done to modernise the IP offices including computerisation of records, creating appropriate place and environment to work efficiently, subscription to major databases for searches. Patent office has digitised most records of Indian patents and now providing access to public for doing searches. This ability has greatly enhanced the value of patent examination. The patent Office's has also made the e-register of patent accessible to public through its website.

All these efforts earned patent offices of India an international cognizance and have been recognised as International Searching Authority (ISA) under PCT. It is anticipated that ISA will start functioning soon.

Facilitating Mechanism

Patent Facilitating Centre (PFC)

Patent Facilitating Centre (PFC) was set up at TIFAC by DST to respond to challenges envisaged after India became part of WTO. PFC was setup with four fold objectives as mentioned below:

1. Introducing patent information as a vital input in the process of promotion of R&D programmes.
2. Providing patenting facilities to scientists and technologists in the country for Indian and foreign patents on a sustained basis.
3. Keeping a watch on development in the area of IPR and make important issues known to policymakers, scientists, industry, etc.
4. Creating awareness and understanding relating to patents and the challenges and opportunities in this area including arranging workshops, seminars, conferences etc.

Now in prevailing techno legal environment Protection and enforcement of intellectual property rights are inevitable in technology management process. With each passing day since we became WTO member, the importance of knowledge about IPR has been rising steeply in India. Management of IPR requires information, skills, knowledge and experience. PFC has been harnessing the creativity of the Indian knowledge generators (scientists and technologists) in order to make India a major intellectual prowess for better presence in the global trade and commerce. PFC aims at new heights through its uniqueness residing in its motto of service 'may I help you' and being a source of comprehensive information – a single window outlet and remaining innovative in the competitive era. In last 15 years PFC has done following work to develop a ecosystem in India as per its mandated role:

Awareness Creation

PFC in its endeavour to create awareness has organized 405 IPR/patent awareness workshops in different parts of the country sensitizing more than 50,000 scientists, technologists, academicians, industrialists and policy makers. These workshops have been spread all over the country, and open to all R&D institutions, industry, academic institutions, government departments, NGOs and individuals.

Facilitation of IP Protection

PFC provides technical and financial support for protecting inventions emanating from academic institutions, schools, R&D institutions and government agencies. PFC has facilitated filing of about 1000 patents in India and abroad. About 150 patents have already been granted. PFC provides all the services including technical services such as patent searches, novelty and inventive assessment free of cost. The ownership of patents rests with institutions generating the inventions. All filing and maintenance work related to patents is carried out by PFC

Patent Information Centres (PICs)

PFC opened 20 patent information centres (PIC) in 20 states. Now this model of PICs is being directly operated by DST and the total no of PIC at present are 23 in 23 states. The PICs are playing the role of a consultant at the State level, by providing guidance in filing of patent application and extending the patent search services and also assisting PFC in conducting workshops in their respective states. In addition to this, PICs are working on registration of potential items such as Geographical Indications (GIs), inclusion of IPR in course curriculum, etc.

Registering Geographical Indications

PFC has taken initiatives to protect many items of traditional excellence which posses that property due to its origin from that geographical area under 'The Geographical Indication of Goods (Registration and Protection) Act'. Registration certificates were issued for six GIs namely Kangra Tea, Kullu Shawls (by PIC-Himachal Pradesh), Muga Silk (by PIC-Assam) and three GIs relating to Malda mangoes (by PIC-West Bengal).

IPR Cells in Universities

PFC has successfully set up IPR Cells in Universities (IPCU) with the help of PICs and universities' vice chancellors. There are 65 IPCUs in existence in 12 states. IPCUs have been created with the aim of guiding university academicians in matters related to IPR like patent searches, IP audit of universities and protecting such inventions through PICs which in turn approach PFC for filing and prosecution of patent applications and maintaining granted patents. The ownership would rest with the universities. The aim is to have IP cell in each University within few years.

Training Programmes

PFC has conducted a number of IPR related advanced training programmes on topics like patent drafting, patent searches, etc from time to time. Recently, DST has entrusted the responsibility of conducting two training programmes on "IPR and WTO issues" for the scientists under its programme National Programme for Training of Scientist and Technologist. Two programme under this scheme are being organized every year now by PFC. PFC conducted a one week advanced training programme for developing countries in July 2010 in association with the NAM S&T Centre. It was represented by representatives from 28 countries. PFC has also done advance training on IPR and technology transfer in India in association with National Institutes of

Health (NIH), US Department of Human and Health Services (HHS) under the project entitled Indo-US Technology Management training and exchange programme awarded by Indo-US Science and Technology Forum. This training programme involved. Such exchange programme also includes internship for Indian IP manager in US especially NIH, USA. Time to time PFC has collaborated with international agencies for advanced training programme.

Women Scientist Scholarship Scheme

Women Scientist Scholarship Scheme (WOS-C) was launched by DST which aimed at utilizing the dormant potential of women scientists qualified in science and technology by giving them opportunities for practicing their knowledge of science and technology in fruitful manner. The scheme aims at training women in the area of IPR especially patents. The women from science background are selected from all India through a written test and interview. They are given training for a period of one year. During the training period they are also paid a stipend of Rs. 12,500 and 17,500 per month depending upon their qualification. The scheme is being operated in four cities at Delhi, Chennai, Kharagpur and Pune. PFC has already trained around 230 women scientists from all over the country in 5 batches. Most of these candidates have found suitable placement in IPR law firms and other IPR agencies. Some of them have become entrepreneurs in the area of IPR as well. Out of these trained women 98 have registered themselves as Patent Agents.

This scheme is unique in the world which gives on job IP training to women who have been away from science due to family reason and wishes to make their career in S&T specially IPR management.

Database on Indian Patents

PFC pioneered three CD-ROM databases on Indian patents; one on patent applications filed in India entitled **Ekaswa-A** and another on patents accepted by the Indian Patent Office entitled **Ekaswa-B,** which were brought out during 98-99. This met a very important need in the country as this was the only digital database on Indian inventions during this period which was available in public domain. The databases cover the data as published in the Gazette of India from January 1, 1995 to December 2004. Later on PFC developed new database named **Ekaswa-C** for patent applications published in Official Journal of Patent Office from January 2005 till June 2008. The **databases** are now available **online** through the PFC website www.indianpatents.org.in free of cost. These databases are also available on CD RoM on nominal charges.

IPR Bulletins

This is the first ever document of its kind in India, available in the public domain on IPR. The monthly IPR bulletin of PFC has been instrumental in spreading awareness about IP related matters especially patents all over the country. The 122 issues of bulletin brought out till now cover analysis of patent applications filed in India in various areas, case studies of granted patents, articles on IPR matters and domestic and international news.

Counseling and Advisory Role

PFC has been instrumental in preparing guidelines of Ministry of Science and Technology for handling intellectual property emanating from projects funded by DST.

PFC has become a national referral point for industry, universities, government agencies, NGOs, foreign embassies and individual scientists, innovators and consultants, for information and advice on IPR related matters, especially upto date patent information. PFC was instrumental in generating critical inputs including conceptual frame work, actual patent data, analysis, etc. for decision making, policy formulation and future planning at the national level in the area of IPR and related matters.

Support Provided by other Ministries

Ministry of MSME has a scheme for encouraging IPR innovations among MSMEs. Under this scheme it gives a grant of Rs. 25,000 on grant of an Indian patent, Rs. 2 Lakhs on grant of a foreign patent and Rs. 1 Lakhs on GI registration to the MSMEs. Similarly, Department of Information Technology reimburses 50 per cent cost on filing a foreign patent to start-ups with a maximum of Rs. 15 Lakhs.

Other agencies paying facilitation role are as under:

1. National Research and Development Corporation (NRDC)
2. National Innovation Foundation (NIF)
3. IP Cells in ministries and department
4. IPR Cells in leading institutions and Universities

Professionals

Patent Attorney/IPR Attorneys

There was time when India had only about 150 registered patent agents and now this figure has passed beyond 1500. Professional with IP expertise has been constantly increasing in this country. Even many professionals are engaged in outsourced IP work also in companies who work as Knowledge Processing Outsourcing (KPOs). Earlier days even for patent attorney only law degree with any graduation was enough but now it is mandatory to have science or engineering degree to be eligible for becoming patent agents. It indicates intention to infuse the professionalism in the system.

Lawyers

Yesteryears were the days when very few takers were there for the IPR as a specialization in law. Beyond agents a special class of lawyer is being developed by various institutes like National Law Schools, NALSAR University etc who would have right acumen to handle IP and are highly specialized on specific area of IP. Here is trend India would have professionals with super specialization in the area of IP like opposition of patents, litigation in courts relating to subject specific patents etc.

Addressing Disputes

Following avenues are available to parties in case of disputes and can take appropriate relief:

1. Appellate Board
2. Arbitration
3. Courts

Education and Training

1. National Law Schools
2. IIT, Kharagpur
3. NALSAR University
4. Optional Course in Law
5. NIPM, Nagpur
6. Women scientist scheme of DST
7. Various other private institutes offers short and long term courses on IPR.

Motivation through Tax Exemption and Duty Waiver

Relaxation in tax has always been the motivation for industry and individuals for investing in instruments which helps in reduction of payable tax. In general government gives various tax benefits for saving like provident funds, health insurance and various other saving instruments. Availing tax benefits are not just tax benefits but R&D work would enable takers to new competitive edge. In the area of IPR also three are some tax benefits given by the Government to Indian innovator and IP holder.

Patentee can Get Waiver of Excise Duty

What kind of Goods will be eligible	*Conditions*	*Exemption Period*
All goods falling under the schedule of the Central Excise Tariff 1985	1. Goods are manufactured by wholly owned Indian Company 2. Totally developed and designed by Indian company 3. Patented in at least two countries amongst USA, Japan and any country of European Union	3 years from the date of commencement of commercial production

Patentee can get benefit of exemption from Drug Price Control Order.

Other than getting tax benefit on obtaining IPR, these exemptions are available one step prior to this stage. Several tax benefits are given for innovation activities at institutional and industry level when efforts are put on R&D activity. These provision are complementary encouragement for creation of IP protectable work. These benefits are listed below:

1. 150 per cent weighted tax deduction on R&D expenditure to companies doing business as listed below:
   (*a*) Biotechnology
   (*b*) Manufacture or production of
      (*i*) Drugs
      (*ii*) Pharmaceuticals
      (*iii*) Electronic equipment
      (*iv*) Computers
      (*v*) Telecommunication equipment
      (*vi*) Chemicals
   (*c*) Manufacture of aircraft and helicopters

   Relating to drugs and pharmaceuticals such deduction also available on following expenditure other than R&D:
   (*a*) Filing of patent application in India
   (*b*) Clinical trials
   (*c*) Obtaining approval of regulatory authorities under any central, state or provincial act.
2. 125 per cent weighted tax deduction for expenses on sponsoring research programme at National Laboratories functioning under ICAR, CSIR, ICMR, DRDO, UGC, DBT, DAE, DOE, IITs and Universities.
3. 40 per cent depreciation allowance in place of 25 per cent in respect of plant and machinery for manufacturing goods based on indigenous technology developed.

Provision of Income Tax Act 1961 (Section 35) may be referred for tax benefits on R&D expenditure and related business.

Tax Liabilities also Need to be Understood when IPR are Licensed

When IP holder/owner in India license or sell the IP in India it would attract central excise, service tax, sales tax and income tax. If IP holder/owner license or sell IP to person or company outside India such transaction would attract service tax, R&D cess and income tax. In third situation if IP holder/owner outside India license or sell IP in India that would attract custom duty, service tax, R&D cess and income tax. It would always be good to get these aspects consulted by tax lawyers before getting in to such agreement to have appropriate clause in rights transfer/share agreement. The tax liability to be well assessed before the transaction, so the appropriate and expected benefits out of such deals remain unaffected.

Important Statistics

8.1 Filing of Patents in India in Last 20 Years

Filing of patents in India in last 20 years is showing in Figure 19.1.

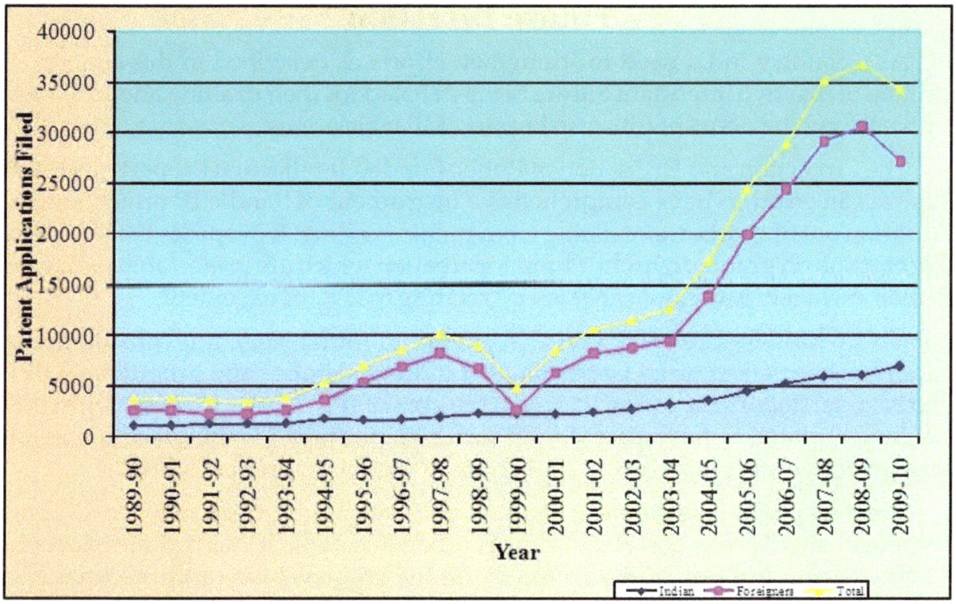

Figure 19.1: Filing of Patents in India in Last 20 Years

Source: Annual report of Patent office 2009-10.

Table 19.2: India's Technology Balance of Payments and receipts 2003-2004 through 2007-2008 in million US$

Year	*Royalty, Copyright License Fees*	*R&D Services*	*Agriculture, Enginee-ring and other Technical Services*	***Total Tech-nology Payment***	*Royalty Copyright License Fees*	*R&D Services*	*Agriculture, Enginee-ring and other Technical Services*	*Total Tech-nology Receipts*
2003-04	444			**444**	32			32
2004-05	712	57	1111	**1880**	71	221	1417	1709
2005-06	729	116	1414	**2259**	191	395	3193	3779
2006-07	1030	201	3025	**4256**	97	760	3457	4314
2007-08	1088	405	3173	**4666**	157	1395	3144	4696

Source: Compilation Sunil Mani in 2009 from RBI data.

Table 19.3: Other IP Applications Filed in India in Last Five Years.

Year	*Industrial Design*	*Trademarks*	*Geographical Indications*	*New Plant Varieties*
2005-06	4949	85699	16	–
2006-07	5521	103419	33	–
2007-08	6402	123514	37	–
2008-09	6557	130172	44	460
2009-10	6092	141943	40	568

Source: Annual Reports of CGPTDM 2009-10 and Annual Report of Protection of Plant Varieties and Farmers Rights Authority 2008-09 and 2009-10.

Future Direction

As a country India need to strengthen efforts as described in this chapter. In addition to that two important bill are being debated for their drafting and redrafting and knocking the doors of future to be part of IP legislations.

The Protection and Utilisation of Public Funded Intellectual Property Bill 2008: This was an effort to have comprehensive instrument of handle IP protection and utilisation out of creation emanating from public funding. It proposes the obligation of protection on grantee of public fund for creation which are protectable under any IP, their ownership, various other issues relating to IPR management.

Utility Model:- discussion on this bill are in initial stage but consultation of matter has been taken up by Department of Industrial Policy and Promotion (DIPP) who is nodal department to enact law in the area of IPR. Under this new legislation it is contemplated to have patent with sorter life for new inventions and possibly lacks inventive step. It is seen as boon to small industries and individuals.

Creating policy instrument at Institutional level: Though now many institutions have policy about protection of IP and sharing benefits. Most of these policies surround benefit sharing but not adequate for fixing the responsibility and essentiality of protecting the creative work under all available IP tools need twist in thinking process while developing these policies.

Conclusion

After having come so far from creation of WTO and sailing in emerging techno legal environment, constant efforts to put India on a right track, in order to take maximum advantages out of it, seem not adequate as yet. Despite various efforts, it appears that a big dent has been made but we seek complete transformation and not just a dent. We still need great efforts on aspects like creating awareness, creating highly skilled human resource by imparting advance training at large scale, building appropriate policy instrument for encouragement for creation/innovation. By addressing these issues at larger level, we would strengthen the ecosystem. Lastly, judicial system has been playing its role very efficiently by disposing IP cases in record time and is thus giving inspirational hope for bright future.

References

Annual Report of TIFAC 2004-05 to 2008-09

Annual Report of Patent Office 2009-10

Annual Reports of Protection of Plant Varieties and Farmers Rights Authority 2008-09 to 2009-10

Damodaran A D 2008 Journal of Intellectual Property Rights, Vol 13 pp 414-423 "Indian patent law in the post TRIPS decade: S&T Policy appraisal

Discussion paper 2011 by Department of Industrial Policy and Promotion (DIPP) on "Utility model"

Saha R 2005 Journal of Intellectual Property Rights Vol 10 pp 369-375 "Capacity building in Management of Intellectual Property Rights – A case of publicly funded institutions"

Saha R (2006) Intellectual Property Rights in NAM and other Developing Countries A compendium on Laws and Policies, pp-40-69

Sunil Mani 2009 working paper "Has India become more innovative since 1991? Analysis of the evidence and some disquieting features"

TIFAC publication (TIFAC-DOC-23) 2007 "Some question and Answer on Patents, Designs, Copyrights, Trademarks, IC Layout Designs, Geographical Indications and Protection of New Plant Varieties."

The Patent Act 1970

Abbreviation

CGPTDM: Controller General of Patents, Designs and Trademarks

CSIR: Council of Scientific and Industrial Research

DAE: Department of Atomic Energy

DIPP-Department of Industrial Policy and Promotion

DoE: Department of Electronics

DRDO: Defence Research and Development Organization

DSIR: Department of Scientific and Industrial Research

DST: Department of Science and Technology

GIPP: Global Institute of Intellectual Property Rights

IC: Integrated Circuit

ICAR: Indian Council of Agriculture Research

ICMR: Indian Council of Medical Research

IDA: International Depository Authority

IGNOU: Indira Gandhi National Open University

IIT: Indian Institute of Technology

IMTECH: Institute of Microbial Technology

INSPIRE: Innovation of Science Pursuit for Inspire Research

IP: Intellectual Property

IPCU: Intellectual Property Cells for University

IPR: Intellectual Property Rights

ISA: International Searching Authority

ISRO: Indian Space Research Organization

MHRD: Ministry of Human Resource and Development

NALSAR: National Academy of Legal Studies and Research University

NBA: National Biodiversity Authority

NIIPM: National Institute of Intellectual Property and Management
PCT: Patent Cooperation Treaty
PFC: Patent Facilitating Centre
R&D: Research and Development
TePP: Technopreneur Promotion Programme
TIFAC: Technology Information Forecasting and Assessment Council
TreMAP: Technology Refinement and Marketing Programme
TRIPS: Trade Related Intellectual Property Rights
WOS-C: Women Scientist Scheme-C
WTO: World Trade Organization

Chapter 20

Status of the Intellectual Property System in Myanmar

Phyo Mon Win
Ministry of Science and Technology
Myanmar
E-mail: phyomonwin@gmail.com

ABSTRACT

At present, we, Myanmar, have drafted four IP laws naming Patents Law, Industrial Designs law, Trademarks and Service marks Law, and Copyright Law. Ministry of Science and Technology, the focal point Ministry for WIPO, and Attorney General Office worked together for drafting the IP laws of Myanmar based on the WIPO Model Law as well as Laws of other IP developed countries.

We had done several discussions on these drafted IP laws with Prime Minister, Ministers, IP lawyers, IP related organizations including NGO, Private Enterprises, Industrial Zones and the concerned from different sectors. The drafted IP laws were sent to WIPO to get the legislative advices on them. WIPO offered Myanmar IP lawyers to come to Geneva for discussion on those laws in May, 2005. In August, 2005, WIPO also sent an expert mission for legislative advices on the drafted Myanmar IP laws.

According to the WTO Press no. 424, all Least-Developed Countries have been given an extension until 1 July 2013 to provide protection for Intellectual Property Rights under TRIPS agreement. Enactment alone is not enough to practice IP system. Its byelaws such as rules, regulations, procedures and other necessary measures (establishing necessary infrastructures (IP office), training officials and staffs, educating the public who will deal with or related to the subject matter etc.) have to be followed up.

Keywords: *IP, Patents, Industrial designs, Trademarks, Service marks, Copyright.*

Introduction

As Union of Myanmar become an ASEAN member on 23rd July, 1997, we need to recognize and respect the protection and enforcement of intellectual property rights in each member state according to the ASEAN Framework Agreement on Intellectual Property Co-operation and also obliged to enhance ASEAN co-operation in intellectual property enforcement and protections. Moreover, Myanmar joined to the World Trade Organization (WTO) as a member on 16 November, 1994 and thereafter also World Intellectual Property organization (WIPO) at 15th May, 2002 as a member which is designed as 176th of member-countries. Consequently, we have the rights as well as obligations that prescribed for its members. Like other members of the WTO, Myanmar has been obliged to abide by and implement the provisions of the Agreement on Trade-Related Aspects of Intellectual Property Rights (TRIPS Agreement). The most comprehensive multilateral agreement on intellectual property concerning with each of the main categories of intellectual property rights: (*i*) copyright and related rights; (*ii*) trademarks including service marks; (*iii*) geographical indications (*iv*) industrial designs; (*v*) patents which include the protection of new varieties of plants; the lay-out designs of integrated circuits; and undisclosed information including trade secrets.

All the laws in Myanmar are in English that enacted in the British era. There had been existing about 300 laws in English after substituting these laws with new laws in Myanmar Language by the successive Governments at the time of the State Peace and Development Council which took over the responsibilities of the State in 1988. There are 173 laws in English still existing even after repeated some of these laws in English and substituted with new ones in Vernacular Language, where necessary.

At the early of 19th century, relevant law as well as official rulings, recognizing and enforcing the principles of IP rights in Myanmar seem to be impressive and had timely developed to a substantial extent. However, most of the old laws become out model and not in harmony with present situations home and abroad due to lack of modification in accordance with the old laws are defunct because of the absent of legal facilities concerned and some had been repealed without substitution.

It does not mean that IP rights are unprotect able at all now in Myanmar. General laws such as Penal Code and some civil laws ensure for protection of the rights of proprietor of the trademark with effective remedies. As a member of the WTO and TRIPS Agreement, we have an obligation to put in place the intellectual laws with the minimum international standards protection. Apart form the TRIPS Agreement, we deeply considered the main objective of ASEAN Framework Agreement that the members must ensure the Intellectual Property Laws and practices in the ASEAN countries will not obstruct the free flow of literary and artistic works, goods and services throughout the ASEAN region but will promote and facilitate such an exchange. As a member of the ASEAN and WIPO, all draft laws existing in Myanmar need to be reviewed and redrafted so as to ensure compliance with international and inter-governmental obligations of Myanmar.

Even though the new full-fledge copyright law has not been promulgated up to now relevant laws which are in some ways cover certain parts of the provisions prescribed in the above mentioned. Agreement has already been enacted. Those laws are the Motion Picture Law, the Television and the Video Law and the Computer Science Development Law, etc.

Among the existing relevant laws that can be enforceable on Intellectual Property Rights, the laws dealing with both criminal and civil action are mentioned in the following:

(*a*) The Myanmar Copyright Act;

(*b*) The Myanmar Merchandise Marks Act;

(*c*) The Myanmar Patents and Designs (Emergency Provision) Acts;

(*d*) The Registration Act;

(*e*) The Penal Code;

(*f*) The Sea Customs Act;

(*g*) The Land Customs Act;

(*h*) The Television and Video Law;

(*i*) The Computer Science Development Law;

(*j*) The Specific Relief Act;

(*k*) The Code of Criminal Procedure;

(*l*) The Code of Civil Procedure.

The Myanmar Copyright Act

Copyright relates to artistic creations such as poem, novels, music, paintings, cinematographic works, etc. The existing Myanmar Copyright Act was promulgated on 24th February, 1914. In fact, the Act contains only 13 Sections with Copyright Act, 1911 Act of the United Kingdom of Great Britain and Ireland, attached there to as the First Schedule. Almost all of the provisions of the Act, with the exception of a few sections that deal with penalties, are provisions, which prescribe how to modify and apply English Act in Myanmar. Myanmar copyright law is not only old but also does not cover TRIPS Agreement.

The term "copyright" means the sole right to produce or reproduce the work or any substantial part thereof in any material form. Copyright is an intangible property and property right intended to protect the right of those who create works of various kinds. The area of copyright includes every production in the literary, scientific and artistic domain. Translation, adaptations, arrangements of music and other alterations of a literary or artistic work shall be protected as original works without prejudice to the Copyright in the original work. It also includes the sole right:

(*a*) to produce, reproduce, perform, or publish any translation of the work;

(*b*) in the case of a dramatic work, to convert it into a novel or other non-dramatic work;

(*c*) in the case of a novel or other non-dramatic work, or of an artistic work, to convert it into a dramatic work, by way of performance in public or otherwise;

(*d*) in the case of literary, dramatic or musical work, to make any record, perforated roll, cinematography film, or other contrivance by means of which the work may by mechanically performed or delivered; and

(*e*) to authorize any such acts as aforesaid.

Copyright shall subsist for the respective term prescribed in the Act, in every original literary, dramatic, musical and artistic work, if it was first published within the Union of Myanmar. There are also definitions for such expressions as literary work, dramatic work, artistic work, work of sculpture, architectural work of art, engraving photography and cinematography.

Infringement of Copyright is doing anything, without the consent of the owner of copyright, the sole right to do which is by the Act conferred on him. Selling, offering for sale, hiring, distributing or exhibiting in public for the purposes of trade, importing for sale or hiring any work shall also constitute an act of infringement of copyright. It shall also be an infringement of copyright if a person for his private profit permits a theatre or other place for entertainment to be used for the performance in public of the work without the consent of the owner of copyright. But there are also exceptions to the infringement of copyright, such as, using any work for private study, research, criticism, review, publishing an address of a political nature delivered at a public meeting, publication in a newspaper of a lecture delivered in public unless such a publication is not prohibited, publishing of printings, drawings, engravings or photographs of a work of art, permanently situate in any public place, reading or recitation in public of any reasonable extract from any published work.

Terms of Copyright

The term for which copyright shall subsist, except as otherwise expressly provided by this act, be the life of the author and a period of fifty years after his death. The exception to this provision is that at any time after the expiration of Twenty-five years from the death of the author of a published work, a person may reproduce it for sale if he has given the notice in writing of his intention to do so and has paid to, or for the benefit of the owner, copyright royalties for all copies of the work, calculated at the rate of ten per cent on the selling price of each copy.

Terms of copyright differs form one another depending on the kind of work created. If it is a work of joint authorship, copyright shall subsist during the life of the author who first dies and for a term of fifty years after his death, or during the life of the author who die last, whichever period is the longer.

In the case of a literary, dramatic, or musical work, or and engraving, in which copyright subsists at the date of the death of the author, but which has not been published, performed or delivered in public, before that date, copyright shall subsist till publication, or performance or delivery in public, whichever may first happen, and for a term of fifty years thereafter.

Copyright in any work which has been prepared by or under the direction or control of any government department, belongs to the Government and subsists for a period of fifty years from the first publication of the work.

In the case of records, perforated rolls, and other contrivances by means of which sound may be mechanically reproduced, the term of copyright shall be fifty years form the making of the original plate from which the contrivance was directly or indirectly derived.

The terms of copyright in photographs is fifty years form making of the original negative form which photograph is directly or indirectly derived.

The author of a work, as a general rule is the first owner of the copyright therein. There are exceptions to this general rule. In the case of engraving, photograph or portrait, where the plate or other original was ordered by some other person and was made for valuable consideration, then the person by whom such plate or other original work was ordered shall, in the absence of any agreement to the contrary, be the first owner of the copyright. And also in the case of the author in the employment of some other person under a contract of service or apprenticeship to make the work in the course of employment by that person, the person who employs the author shall, in the absence of any agreement to the contrary, be the first owner of the copyright.

If copyright in any work has been infringed, the owner of the copyright is entitled to all remedies by way of injunction or interdict, damages or accounts. In those cases the Specific Relief Act or the Law of Tort may be applied, depending on the nature of the case. However, an action in respect of infringement of copyright shall not be commences after the expiration of three years next after the infringement.

In addition to civil remedy, the owner of the copyright may resource to the court for taking criminal action against the infringer under the Myanmar Copyright Act. Chapter III of the Acts prescribes penalties, though they may not be supposed to be sufficient punishment for the offense in the present day situation, for the following offences:

(*a*) Knowingly making for sale or hire any infringing copy of a work in which copyright subsists;

(*b*) Knowingly selling or letting for hire or by way of trade exposes or offers for sale or hire, any infringing copy of any such work;

(*c*) Knowingly distributing infringing copies of any such work, either for the purpose of trade or to such an extent as to affect prejudicially the owner of the copyright;

(*d*) Knowingly by way of trade exhibits in public any infringing copy of any such work;

(*e*) Knowingly importing for sale or hire into the Union of Myanmar any infringing copy of any such work;

(*f*) Knowingly making or having in possession any plate for the purpose of making infringing copies of any work in which copyright subsists;

(*g*) Knowingly and for his private profit causing any such work to be performed in public without the consent of the owner of the copyright.

For infringement of copyright in television and video, there is a specific law entitled" The Television and Video Law" which prescribes penalties for certain offenses. Under section 33 of that Law, the offender of such offenses as copying, distributing, hiring or exhibiting for commercial purpose a censor-certified video tape, without the permission of the license holder, may be punished with imprisonment extending up to three years or with fine extending up to Kyats 100,000 or with both. In these cases, as the penalty is supposed to be effective and deterrent and the trial of offenders is usually expeditious, the owner of the copyright may not have resource on civil action. However the fact that the infringer has been punished under the relevant penal law shall not be a bar to civil litigation.

In addition, every suit or other civil proceeding regarding infringement of copyright shall be instituted and tried at the Supreme Court or at the Court of District Judge under Section 13 of the Myanmar Copyright Act.

The Court before which any of the foregoing is tried may, whether the alleged offender is convicted or not, order that all copies of the work or all plates in the possession of the alleged offender, which appear to it to be infringing copies, or plates for the purpose of making infringing copies be destroyed or delivered up to the owner of the copyright, or otherwise dealt with as the Court may think fit.

Although the act is still in existence, very few cases have been brought before the Court. The main reason for scarcity of cases relating to copyright is that the persons concerned with such disputes are reluctant to go to Courts of Law, which is a distinctive personal trait of most of Myanmar people. Almost all of such disputes are settled amicably by negotiations or conciliations; sometimes with the intervention of certain influential persons or bodies such as Literary and Journalist Organization, Motion Picture Organization, Music Organization, Thabin (Performing Arts) Organization and Artists and Artisans Organization. As the Act is still in force, any person whose rights there under is infringed may seek protection in accordance with the Act.

Trademarks

In Myanmar, there is no specific statutory law on trademarks nor is there any statutory provision regarding registration of trademarks. This was reflected in a 1951 ruling of the then Supreme Court of the Union. The then Chief Justice observed.

In Burma, we do not have any Act like the English Trademarks Registration Act, 1875, the English Trademarks Act, 1905, or the Indian Trademarks Act, 1940. So there is no system for registration of trademarks and there is no statutory title to trademarks.

There being no statutory law on trademarks, a trademark, as understood in Myanmar Civil Law, is not defined. However, there exists the definition of a trademark in the Penal Code. Section 478 of the Code defines a trademark as "a mark used for denoting that goods are manufactured or marketed by a particular person". According to this definition, even words can constitute a trademark. Thus, it should be noted that, in Myanmar, the word "Trademark" can be used in a very wide sense, there being no Trademarks Act. The question of "what is a trademark" has to be decided on the basis of the rulings of the highest Court of Law, which again based on the English Common Law.

In order to be a trademark, as defined in Section 478 of the Penal Code, a mark must be distinctive so as to distinguish the goods of the proprietor of a trademark from those of other persons. "Distinctive" means distinguishing a particular person's goods from somebody else's-not a quality attributed to the particular article. If a mark merely describes the quality or origin of the article, such a descriptive mark is obviously not capable of distinguishing the goods of one maker from those of others. It has been held in one case that the style or the get-up of the boxes or packages in which goods are retailed does not constitute a trademark as defined in Section 478 of the Panel Code.

It does not really matter whether a particular brand or label used by a particular manufacturer is registered as a trademark in the strict legal sense. Thus, the manufacturer may use a particular color combination only, without any particular mark, and anyone who imitates that particular color combination on his goods may be prosecuted or sued or both actions may be taken against him. In Myanmar, many manufacturers usually register the entire label attached on the goods, although within that label they may have a particular device which may be called a trademark. In such a case, the whole label will be a trademark and not the particular device in it.

Registration Act

In Myanmar although there is no operative Trademark Registration Act nor is there any provision for a statutory title to a trademark, a firm practice has developed for the evidentiary assertion of the rights of ownership in a trademark. The trademark registration system in Myanmar has been operating under Direction 13 of the Registration Act which explicitly provides that trademarks may be registered under Section 18(F) of the Registration Act. Under that Direction, a trademark may be registered with the Office of Registration of Deeds by means of a declaration, which is a solemn statement of facts made by the owner of the trademark and usually attested by a Notary Public or a Magistrate or a Judicial Officer.

Although the mere fact of registration is not conclusive proof of the ownership or user of the trademark, nonetheless it may be ***prima facie*** evidence of ownership or user and the ability to produce a registered document may be of some help in a criminal or civil proceedings. The registration, as such, is usually followed by an advertisement or announcement in daily newspapers, stating that any fraudulent imitation or unauthorized use of or infringement of the said trademark will be dealt with according to law.

In the process of registering a trademark, the owner of it shall mention the following facts as admissions in the document or declaration submitted for registration, that:

(*a*) The said trademark is being used for the type of goods which he manufactured and distributed for sale;

(*b*) The said trademark has been created by him out of his own idea;

(*c*) The said trademark is neither an imitation nor a fraudulent copy of a trademark owned by any other person;

(*d*) In his knowledge, no person has manufactured and distributed any goods of the same type using the said trademark.

For registration of the document of declaration, the charge for the revenue stamp is merely Kyat 25/- and the fee collected for registration is merely Six Kyats. Foreign trademarks, if certified by a Notary Public may also be registered under the same conditions and with the same payments. The Date of registration shall not also amount to the commencement of one's right like in some other countries, instead only the time of use of the mark. There is no provision in the Registration Act for the validity period or renewal time, therefore registration once stays perpetual, no need for renewal. However, in case of any material changes supposing proprietor's name, device, address or covered goods services and so on to the registered declaration, it should have to be registered afresh. Even so done, the owner can maintain the earliest registration date of the mark over subsequent registration dates. Lack of specific trademark law, there are isolated provisions in various status, which are happened in touch with relevant matters to trademark and its ambiences. In addition to the civil litigation, criminal action under the Penal Code also is possible against a person using false trade mark or counterfeiting a trade mark or making or possessing any instrument for counterfeiting a trade mark, or selling goods marked with a counterfeiting trade mark.

In the absence of a statutory trademark law in Myanmar, plaintiffs were for many years restricted to bringing common law suits for "passing off". However, in 1968 the Supreme Court held that a suit for infringement of a trademark is maintainable in Myanmar.

Remedies for Infringement

According to a ruling in 1939 of the late High Court of Yangon, no civil action was maintainable for infringement of trade mark rights, and the only action maintainable in civil law was an action for passing-off, which is an old common law remedy, there being no Trademarks Act in Myanmar. This concept was accepted and followed by the Courts of Myanmar up to 1968. But in 1968, the correctness of this ruling was challenged in case before the then Chief Court, and the learned Judge dissented from the view mentioned in the former case and overruled the former decision and held that an action for infringement of the trademark is maintainable under Section 54 of the Specific Relief Act, which clearly authorizes the Courts in Myanmar to grant perpetual injunctions in infringement suits. Thus, since the passing of judgment in that case, a person whose right to a trademark has been infringed can maintain an action for infringement of the trademark as well as action for passing-off.

In an infringement action, the question of fraud or probability of deception is immaterial, and all that the plaintiff has to prove is that he is the true owner of the trademark and that the defendant is using a get-up which is either the same or similar to his trademark.

In a passing-off action, the proprietor first has to prove that he has used the mark and has acquired a reputation in it in the business concerned. He then has to prove that what the defended has been going is to cause the defendant's goods to be confused with the plaintiff's goods with the result that customers are led to believe that they are obtaining goods of the plaintiff when in fact they are obtaining goods of the defendant.

Through the rules of procedure and evidence are different, the relief given by the Courts in suits for passing-off are the same as in suits for trademark infringement. The relief obtainable in these actions is injunctions, damages and other ancillary relief amongst which an injunction is in the most effective as it is a specific order of the Court directing the defendant to refrain from the act of infringement or passing-off action.

In addition to civil litigation, a person who has established the right to a trademark may take action against anyone for the following offenses:

(*a*) Using a false trademark, punishable with imprisonment up to three years or with fine or with both;

(*b*) Counterfeiting a trademark, punishable with imprisonment up to three years or with a fine or with both;

(*c*) Making or possessing any instrument for counterfeiting a trademark, punishable with imprisonment up to three years, or with a fine or with both;

(*d*) Selling goods marked with a counterfeit trademark, punishable with imprisonment up to one year or with fine a or with both.

Besides the above-mentioned provision in the Penal Code, there exists the Myanmar Merchandise Act, which supplements relevant provision in the Penal Code. The Act empowers the Court, ***inter alias***, to confiscate all goods and things by means of, or in relation to, which the offense has been committed.

The Myanmar Merchandise Marks Act

According to the Myanmar Merchandise Marks Act, the term "trade mark" has the meaning assigned to that expression in section 478 of the Penal Code and that of trade description, statement or other indication, direct or indirect, as to any goods being the subject of an existing patent, privilege or copyright.

In section 6 of the said Act, penalty for applying a false trade description is prescribed. In such manner, penalties for selling goods to which a false trade description is applied and unintentional contravention of the law relating to marks and descriptions are prescribed in Sections 7 and 8 respectively.

Regarding forfeiture of goods, the provision is provided in Section 9. Section 22 provides for the punishment of abetment.

The Patents and Designs (Emergency Provisions) Act

The legislature in Myanmar has enacted a law to protect inventions, a category of industrial property. Burma Patents and Design Act, 1939, was first introduced to Myanmar after separation, in 1937, of the then Burma from India. Then the Myanmar Patents and Designs Act of 1945 was promulgated but was not brought into force till the law was repealed in 1993. However, the Burma Patents and Designs (Emergency Provisions) Act, 1946, which came into force retroactively on July 1, 1941 is still in force with only two sections in it. Section 2 of this Emergency Provisions Act provides that, until the Burma Patents and Design Acts, 1945, comes into operation, the Indian Patents and Designs Act, 1911, shall continue to have effect in Burma. In fact, both Myanmar Patents and Designs (Emergency Provision) Act, 1946 and Myanmar Patents and Designs Act, 1945 are defunct so that it may be said that there is presently no law or at least any law in operation on patents and designs. In these circumstances the Attorney General's Office has drafted new laws on Patents and Industrial Designs and forwarded to the Ministry of Science and Technology for comments and necessary action.

The Penal Code

Myanmar does not lack bringing into justice people who have broken criminal law. The Penal Code of Myanmar is the substantive law for this purpose.

Regarding trade, property and other marks, the definitions of trade make and property mark are prescribed in Sections 478 and 479 respectively, and using a false trade mark and property mark are prescribed in the following two sections–Sections 480 and 481 respectively. Punishment sections followed suit covering Sections 482 through 489. The following are offences as regards the said marks:

(*a*) Using a false trade mark or property mark

(*b*) Counterfeiting a trade mark or property mark used by another;

(*c*) Counterfeiting a mark used by a public servant;

(*d*) Making or possessing of any instrument for counterfeiting a trademark or property mark;

(*e*) Selling goods marked with a counterfeit trademark or property mark;

(*f*) Making a false mark upon any receptacle containing goods;

(*g*) Making use of any such false mark; and

(*h*) Tempering with property mark with intent to cause injury.

The Sea Customs Act and Land Customs Act

The Sea Customs Act prohibits export or import by land or sea of goods having applied a counterfeit trade mark. The same applies in the area of trade and commerce. Under sections 18 and 19 of the Sea Customs Act, the importation into the country by land or sea of goods having applied thereto a counterfeit trademark within the meaning of the Penal Code or a false trade-description within the meaning of the Myanmar Merchandise Marks Acts and goods made or produced beyond the limits of the Union of Myanmar and having applied thereto any name or trademark being, or purporting to be, the name or trademark of any person who is a manufacturer, dealer or trader in the Union of Myanmar is prohibited. Such goods, imported in contravention of this provision shall be liable to detention and confiscation, and in addition any person involved in any such offence is liable to pay fine under sections 19A and 167 respectively.

In section 170 and 171, the Customs Officer is authorized to stop and search any person vessels and vehicles for goods on reasonable suspicion.

In section 9 of the Land Customs Act, it is prescribed that the provisions of the Sea Customs Act including the above-mentioned ones are applicable.

The Science and Technology Development Law

In 1994, a new law entitled "The Science and Technology Development Law" has been enacted to cope with the changing situation and development in the market-oriented economy and to enhance its further development. The law provides that every contact for transfer of technology must be registered and failing which it will not have a right to institute a suit based on it. The objectives of the Law are prescribed in Section 3, which states as follows:

(*a*) To development science and technology and promote industrial production so as to contribute towards the National Economic Development Plan;

(*b*) To carry out research and development for the increased extraction and utilization of domestic raw materials and the promotion of industrial production based on modern science and technology;

(*c*) To effect technology transfer for the promotion of production processes and the improvement of the quality of goods;

(*d*) To nurture experts required for the development of science and technology and for research and development and to improve their qualifications;

(*e*) To communicate and cooperate with domestic and foreign research institutes and organizations for the development of science and technology and research and development;

(*f*) To honor and grant appropriate benefits to outstanding experts and inventors in the field of science and technology.

National Council

Section 4 of the Law prescribes for the formation of the "National Council for Science and Technology Department" which is a high-level body headed by a responsible person not below than the rank of a cabinet member and consisting of ministers or deputy minister from the relevant ministries, heads of government departments and organization mainly responsible for the development of science and technology, and appropriate experts.

The duties and powers of the National Council are, ***inter alias***, to lay down policies for the successful implementation of the objectives of the law; to provide guidance and supervise the scientific and technological development and research development activities, conducted by government departments and organizations, and to promote and assist the advancement of science and technology works and research and development activities by non-governmental organizations, cooperative societies and private experts.

Rules for Transfer of Technology

Another important subject dealt with by the Law is Technology transfer. According to the provision of the Law, a person or an organization that has conducted research and successfully developed a new technology may transfer the right to use it to any person or organization upon payment of a fee or free of charge under a licensing contract. The manner and conditions relating to the export and import of technology shall be determined by the Government. The law also prescribed some salient terms and conditions to be contained in contracts for technology transfer. Every contract for transfer of technology must be registered with the Registry established by the Department. There shall be no right to institute a suit based upon a contract which is not registered. However, the law provides that the provisions relating to technology transfer shall not apply to the transfer of a right to patent on design, for which separate specific laws are expected to be introduced in the very near future. But transfer of computer technology, according to Section 40 of the Computer Science Development Law promulgated in 1996, are to be made in accordance with the provisions of the Science and Technology Development Law, as have been mentioned above.

The Television and Video Law

For infringement of copyright in television and video, there is a specific Law entitled "The Television and Video Law of 1996" which prescribes penalties for certain offences. Under section 33 of that Law, the offender of such offences as copying, distributing, hiring or exhibiting for commercial purpose a censor-certificated video tape, without the permission of the license holder, may be punished with imprisonment extending up to three years or with fine extending up to Kyats 100,000 or with both. In these cases, as the penalty given is effective and deterrent and the trial of offenders is usually expeditious, the owner of the copyright does not wish to have recourse to civil action. However, the fact that infringer has been punished under the relevant penal law is no bar to civil litigation.

The Computer Science Development Law

As regards protection of software piracy no protection can be sought due to the lack of advanced technology. However, section 36 of the Computer Science Development Law, enacted in 1996 penalizes for imports or exports of any type of computer software or any information prescribed by the Myanmar Computer Science Development Council under Section 6(g) of the said Law.

The Specific Relief Act

Under section 42 of the Specific Relief Act, any person entitled to any rights as to any property, including intellectual property, may institute a suit against any person denying, or interested to deny, his title to such right. The Court may in its discretion make therein a declaration that he is so entitled. In addition, the plaintiff, under section 54 of the said Act, may also claim a decree for a perpetual injunction. The object of granting perpetual injunction is to refrain from invading or threatening to invade the plaintiff's enjoyment of the intellectual property rights. Perpetual injunction can only be granted by a decree made after hearing on the merit of the suit.

The Code of Criminal Procedure

In Myanmar, no specific procedures exist for search, seizure and disposal of Intellectual Property. However, the general provisions of the Criminal Procedure Code relating to search, seizure and disposal of property can be applied. Section 101 to Section 103 prescribes that search shall be done in the presence of at least two witnesses, who are respectable inhabitants of the locality, and that the occupier of the place searched may attend during the search, and they are required to sign the list prepared in token of the correctness thereof.

The property found in the course of search shall become an exhibit of the proceeding and the Court has authority to dispose of the property in accordance with sections 516-A and 517 of the Criminal Procedure Code. Section 516-A provides that the Court may make an order, as it thinks fit, for the custody or disposal of property during the pending of a trial in certain cases. Section 517 states when an inquiry or a trial in any Criminal Court is concluded, the Court may make such order as it thinks fit for the disposal, by destruction, confiscation, or delivery to any person claiming to be entitled to possession thereof, or otherwise, of any property or document produced before it or in its custody, or regarding which any offence to have been committed, or which has been used for the commission of any offence.

The Code of Civil Procedure

Preliminary Measures before Hearing of the Suit

The Civil Law of Myanmar also plays a role in this subject. There is the Civil Procedure Code to oversee this.

In the settlement of disputes relating to Intellectual Property infringements, Civil Procedure Code plays an important role in Myanmar. Under the said Code there are some effective procedures for the right decision and speedy disposal of the suits.

Where issues both of law and of fact arise in the same suit, and where the Court is of opinion that the case may be disposed of on the issues of law only, it shall try those issues first. Where the parties to a suit agree as to question of fact or of law to be decided between them, they may state the agreement in the form of an issue, and enter into an agreement in writing. Then the Court if it is satisfied with the agreement produced by the parties, may proceed to record and try the issues and state its finding or decision thereon, in the same manner as if the issues are framed by it; and may pronounced judgment according to the terms of the agreement; and upon the judgment so pronounced, a decree shall follow.

Any party to a suit may give notice, by his pleading or otherwise in writing that he admits the truth of the whole or any part of case of any other party; and also either party, may call upon the other party to admit any document or any fact stated in the pleadings. In addition, the plaintiff or defendant by leave of the Court may deliver interrogatories in writing for the examination of the opposite party; and the opponent shall have to answer the interrogatories by affidavit.

Under Order 8, Rule 5 of the Civil Procedure Code, every allegation of fact in the plaint, if it is not denied specifically or by necessary implication, or stated not to be admitted, in the written statement of the defendant, shall be deemed to be admitted, except as against a person under disability.

Under Order 12, Rule 6 of the Civil Procedure Code, any party may at any stage of a suit, where admissions of the facts has been made, either on the pleadings or otherwise, apply to the Court for a judgment, degree or order which he may be entitled to, without waiting for the determination of any other question between the parties; and the Court may either upon such application or upon its own motion, give such judgment or make such degree or order as it may thick just. The degree or order passed under this rule may be executed at any time notwithstanding that other questions between the parties still remain to be decided in the case.

Therefore according to the above mentioned procedures of the Code, a suit may be disposed of in the preliminary stage within a short period before taking any evidence.

The above procedures are effective remedies that are prescribed in the Code of Civil Procedure for speedy disposal of the suit, and to save the time and expense of the parties.

Procedure in the Original Suit

In the absence of the specific intellectual property law in Myanmar, under Section 9 of the Code of Civil Procedure, the Courts have jurisdiction to try all suits of a civil nature excepting suits of which cognizance is either expressly or impliedly barred. Under section 151 of the said Code, where there are no relevant provisions in the Code, then the Judges may also exercise the inherent power as may be necessary for the ends of justice.

To be in conformity with the Civil Procedure Code, an action has to be commenced with a plaint by the owner of the Intellectual Property Rights as plaintiff, and the alleged infringer as defendant.

The defendant has to submit a written statement if he does not admit the allegations in the plaint. After that, the Court has to frame the issues on which the right decision of the case may be made. Then, the witnesses of the plaintiff and the defendant shall be examined on oath. After examination of the parties and their respective witnesses, the judgment and decree have to be passed by the court.

Temporary Relief's before Judgment

Under Order 39, Rules 1 and 2 of the Code of the Civil Procedure, the Court may grant temporary injunction in every suit during the trail of the case when the following conditions are proved:

1. That any property in dispute in a suit is in danger of being wasted, damaged or alienated by any party to the suit, or
2. That the defendant threatens, or intends, to remove or dispose of his property with a view to defraud the opposite party to the suit.

In addition to the provision mentioned above, in any suit for restraining the defendant from committing a breach of injury of any kind, whether compensation is claimed in the suit or not, the plaintiff may, at any time after the commencement of the suit, and either before or after judgment, apply to the Court for a temporary injunction to restrain the defendant from committing the breach of injury complained of, relating to the property or right which is the subject matter of the suit.

In case of disobedience or branch of any terms of the temporary injunction, or any interim orders mentioned above, the Court may order the property of the person guilty of such disobedience or breach to be attached, and may also order such person to be detained in the civil prison, for a term not exceeding six months, unless in the meantime the Court directs his release. Also the property attached under such interlocutory orders may be sold, and out of the sale proceeds, the Court may award such compensation as it think fit, and shell pay the balance to the party entitled thereto.

Regarding infringement cases taken by civil action against the infringer, if the Court is satisfied by an affidavit or otherwise, that the infringer or the defendant is about to abscond, or is about to remove or dispose of the whole or any part of his property, from the local limits of its jurisdiction, the Court may either issue a warrant to arrest the defendant and bring him before it, to show cause why he should not furnish security for his appearance, under Order 38, Rule 1 of the Civil Procedure Code, or may direct a conditional attachment of such property under Order 38, Rule 5 and 6 of the same Code.

By virtue of Order 39, Rule 7 of the Civil Procedure Code, the Court may on the application of any party to a suit, and on such terms as it thinks fit, make an interim order for the detention, preservation or inspection of any property which is the subject matter of such suit, and the Court may authorize any person to enter upon or into any land or building in the possession of any party to such suit. The Court may also authorize any samples to be taken or any observation to be made, or experiment to be tried, which may be necessary for the obtaining full information or evidence.

Under Order 40, Rule 1 of the Civil Procedure Code, where it appears to be just and convenient, the Court may appoint a receiver of any property, whether before or after decree: and may commit the same to the possession, custody, or management of the receiver. The Court may also confer upon the receiver all such powers, as to bringing and defending suits; and for the realization, management, protection, preservation, improvement, collection and disposal of mesne profits of the property, according to the direction of the Court.

Procedure Relating to Decree and its Enforcement

In some of the cases, if necessary, the Court may before passing the final decree, pass a preliminary decree, and may appoint a commissioner, to inquire and to take accounts or damages suffered by the decree-holder. After obtaining the commissioner's report and hearing the arguments of the parties or their pleaders, the Court has to pass final decree. For the enforcement of the decree, an execution case shall be followed in continuation to the original suit, to fulfill the rights and reliefs of the decree holder.

In any proceeding of execution of decree, in which the executing Court deems a local investigation is necessary or proper for the elucidating any matter in dispute, or for ascertaining the market value of any property, or the amount of any mean profits or damages, or annual net profits, it may issue a commission to such person as it think fit, directing him to make such investigation and to report thereon to the Court, under Order 26, Rule 1 of the Civil Procedure Code.

The Court executing a decree, may issue a warrant for the destruction of the intellectual properties involved in infringement, and of the trade marks in possession of the infringer or judgment debtor, in addition to the orders for execution of the main relief passed under relevant Laws and Acts, which are drawn up in the decree. These are brief accounts on some of the procedures relating to the execution of a decree, under the respective Rules of Order 21 of the Civil Procedure Code.

Injunction

As injunction is the most effective remedy obtainable on trademark and copyright infringement suits, the owner can seek preventive relief granted at the discretion of the Court by injunction, temporary or perpetual. Temporary injunctions are to continue until a specified time, or until the further order of the Court. They may be granted at any period of suit, and are regulated by Order 39, Rule 1 of the Civil Procedure Code. Under Section 54 of the Specific Relief Act, perpetual injunction can only be granted by the decree made after hearing and upon the merits of the suit. The defendant is perpetually refrained from invading or threatening to invade the plaintiff's enjoyment or property.

Expert Witnesses

Section 45 of the Evidence Act states that the opinions of experts skilled in the subject by a special course of study, practice or training are relevant to a point of (a) foreign law or (b) science or art or (c) the identity of handwriting or finger impressions. When the Court has to call an opinion upon a point relating to Intellectual Property Rights, the opinions of experts, specially skilled in the subject are admissible under Section 45 of the Evidence Act.

Thus, both parties may call for expert witnesses to establish their cases. Their depositions will be the evidence that will be weighed by the judge. There is no expert advisor appointed for the Judge according to the existing procedural laws in Myanmar.

Role of Judges

Under the Judiciary Law, the Supreme Court and different levels of the Courts are formed to adjudicate on original criminal and civil cases as well as appeal and revision. The Supreme Court is the highest judicial organ of the State and delegates its functions and powers to the Judges of the different levels of the Court under its supervision.

Although there is no specific Court to try the cases on intellectual property rights, Judges of different levels of the Court, have jurisdiction in all the suits of such nature, under Section 9 of the Code of Civil Procedure excepting suits of which jurisdiction is either expressly or impliedly barred. In addition, all the Judges also have jurisdiction to exercise the inherent power invested to them, where there are no specific provision in the Code of Civil Procedure.

In addition, all the Judges, according to their jurisdiction, have power to take criminal action for infringing the intellectual properties rights under the provisions of the Penal Code and other special laws. The properties involved in the offence may also be searched, detained and confiscated by the Judges under sections 101 to 103, 516A and 517 of the Code of Criminal Procedures.

Role of Prosetuators

Prosecutors, also known as Law Officers in Myanmar, act effectively as Ex-Officio Public Prosecutor who plays a critical role in the administration of justice. Under section 2 of the Attorney General Law, the Attorney General and the Deputy Attorney General are appointed by the supreme authority of the State and duties and powers are conferred on them.

Under section 4 of the said Law, the Attorney General is authorized to form the Attorney General's Office and different levels of Law Offices all over the Country as may be necessary. He is also authorized to appear at the Supreme Court and act as the Government Attorney in criminal and civil cases of the Government and instructing Law Officers for such appearances.

Under section 3 of the Attorney General Law, the Attorney General is assigned duties among which are:

(*a*) Tendering legal advice when so requested by the State Peace and Development Council, the Government or any Ministry;

(*b*) Filing necessary appeal or revision to the Supreme Court in respect of adjudication of Courts at different levels that are not in accordance with law;

(*c*) Scrutinizing, drafting, translating and revising laws to conform to the times; and

(*d*) Making orders, directives, procedures and manuals as may be necessary.

Subject to powers conferred on him the Attorney General has issued specific instructions to the Law Officers assigned to different levels of Law offices to perform the prescribed functions and duties among which are the followings:

(*a*) Tendering legal advice to government departments and organizations at respective levels;

(*b*) Appearing at the Court as a Public Prosecutor at the trial of a person who has violated criminal law:

(*c*) Acting and pleading for the State in civil cases where the State is plaintiff or defendant;

(*d*) Tendering advice in cases which are to be sent up to the Court to be in accordance with law;

(*e*) Examining any complaints lawfully made by a citizen who is said to be deprived of his rights and privileges and taking necessary measures in accordance with the directive of the Attorney General; and

(*f*) Submitting to the Attorney General judgments, orders and decision of the judicial Courts which are found to be inconsistent with law.

Regarding the role of the Law Officers in the institution of a criminal case, the stages in practical terms are as follows:

After the complaint is filed at the police station, the investigation process such as arresting the suspect(s) and the accused, seizing the exhibits, examination of witnesses and framing of the case is carried out by the police force. Then, the documents taken during the said process called Police Papers are sent to the relevant Law Office whereupon the Law Officers have to scrutinize the Papers from comprehensive legal point of view. In a situation where there is no sufficient ground for instituting of the case at the Court, the Law Officer is authorized to close the case or to give legal advice for further investigation. In different situation, the respective police force is advised to submit the case to the Court for trial giving detailed instructions for proper institution of the proceedings.

In the trial of the case the Law Officer is responsible to appear at the Court as the Public Prosecutor. The trial is usually initiated by examining the prosecution witnesses conducted by the Law Officers. Given the evidences recorded in the proceedings of the case, the charge is framed against the accused or discharged him. At this stage, if charge or discharge is not in line with the law, the Law Officer is empowered to take appropriate action for revision to the higher Court.

After charge in framed against the accused, the defense witnesses are examined by the Defence Attorney and cross-examined by the Law Officer. Afterwards, the judgment is passed by the Court. Hereupon, the Law officer is authorized to take appropriate action such as appeal against acquittal or revision for enhancement of punishment.

In addition, under Section 3 (c) of the Attorney General Law, the Attorney General under whose instruction the Law officers of different levels of Law offices perform the duty conferred on them on give legal protection to the people in order to enable them to fully enjoy their legal rights and privileges.

Conclusions

The Government and the people of Myanmar are endeavoring and exerting their best efforts for all-round development of the country in pursuance of political, economic and social objectives, set by the State Peace and Development Council. For a developing country, investment in every sector of its economy, both domestic and foreign, is a key factor to promote its economic development, which forms the basis for all other developments. Union of Myanmar has invited any participation in terms of technical know-how as well as investments from sources inside the country and abroad in order to obtain the proper evolution of the market-oriented economic system and development of the economy for her country and people. To encourage and promote investment, there must be transparent and effective laws consistent with the prevailing economic situation. To be in line with the economic system, the State Legislature has taken positive action to facilitate legal framework for smooth running of businesses, including foreign trade sector and other trade-related aspects.

The protection to intellectual property rights is given great attention as an incentive and encouragement in promoting investment and economic development.

The effective and valid protection of such rights can only be obtained by way of legislation. Laws on intellectual property must be comprehensive and transparent and must contain sufficient provision to ensure effective mechanisms to convince investors that their rights are safeguarded and protested. In view of this fact, the State Peace and Development Council enacted the Foreign Investment Laws in less than three months after it assumed power. The Myanmar Citizens Investment Law was also enacted at that time. In both laws, it is provided that rights to trademarks, patents and technical know-how shall be evaluated and included in investment.

In Myanmar, Intellectual property system is not well-developed. Only after the adoption of the market-oriented economic system, the importance of this subject has been realized and particular attention has been focused on it. In spite of the fact that there exists no specific law regarding the intellectual property rights in Myanmar, infringements on the said rights can be remedied under the existing laws as mentioned above. Existing laws relating to intellectual property are obsolete and defunct and also do not cope with the current trends and developments in the field of intellectual property.

Myanmar joined the World Trade Organization (WTO) as a member, and thereafter also abides by the provisions of the relevant treaties of such organization, in particular the TRIPS Agreement and the ASEAN Frame-work Agreement on Intellectual Property. As a result, all draft laws have to be reviewed and redrafted so as to ensure compliance with international and inter-governmental obligations of Myanmar. Establishing the new Ministry of Science and Technology and designating it as the focal point to deal with intellectual property matters will help expedite the outcome of the new laws. The Supreme Court and the different levels of the Courts of Laws has also been exercising the judicial power conferred them in proper way taking into consideration of the intellectual property rights enshrined in international conventions and agreements including TRIPS Agreement.

The TRIPS Agreement contains an entire part devoted to enforcement. The aim of that part is to ensure that the provisions on protection are enforced properly. It contains both general obligations and special obligations in this respect. It also gives a particularly prominent place of the enforcement of intellectual property rights and contains very detailed provisions in that respect. Being fully aware of the fact that, the enforcement mechanism of the foregoing laws does not cope with the norms and standard provided in the TRIPS Agreement, Myanmar has been endeavoring for drafting and in enacting new laws on intellectual property. The State Peace and Development Council has assigned the Attorney General's Office has been working on successive draft laws for the protection of IPRs in line with provisions of the TRIPS Agreement as well as the ASEAN Framework Agreement on Intellectual Property.

The protection and enforcement of intellectual property rights should contribute to the promotion of technological innovation, to transfer and disseminate of technology, to mutual advantage of producers and users of technological knowledge and in means conducive to social and economic welfare and balance of rights and obligations. At present, the final drafts of such laws were finished and submitted through the government to the State Peace and Development Council for Legislation.

Enactment alone however hardly fills the vacuum in IP field. Its byelaws such as rules, procedures and others necessary measures have to be followed up, so also establishing necessary infrastructures, training officials and staffs as well as educating the public who will deal with or relate to the subject matter. Such tremendous tasks shall have to be done not only ahead the year 2005 but also later on and on. It is also hoped that this workshop, which denotes the goodwill WIPO has, further rendered to our country, will be of much help to our works. We also hope that in the future there will be more contact and close cooperation between Myanmar and WIPO.

References

U Kyaw Zay Ya, 2003. The Myanmar Copyright Act.

Dr Cho Cho San, Assistant Lecturer, University of Computer Studies, Yangon, Myanmar, 2002. Current Situations of the Intellectual Property Rights System in Myanmar.

U Than Maung, 2002. A comparison on Trade Mark system in Myanmar to Singapore.

Mr. Kyaw Lin, Diputy Director, Supreme Court; Mr. Tin Win, Diputy Director, Supreme Court, Mr Kyaw Sein, Diputy Director, Attorney General's Office, 27 November, 1999. Overview of Laws Relating to Intellectual Property Rights and its Enforcement in Myanmar.

Mr. Kyaw Aye, Director General, Office of the Attorney General of Myanmar, Intellectual Property Legislation and Administration in Myanmar.

Mr Than Kyaw, Deputy Director, International and ASEAN Legal Affairs Division, Office of the Attorney General Yangon, Myanmar, A Report on Enforcement of Copyright and Neighboring Rights in Myanmar",

Mr Tun Yee, Assistant Director, Office of the Attorney General, Current Situation of the Enhancement of Intellectual Property Rights in Myanmar.

Mr Than Maung, Advocate; Former Staff Officer of the Attorney General Office, Protection of Trademark Rights in Myanmar.

The Myanmar Merchandise Marks Act, 1889. Myanmar Code Vol V.

The Specific Relief Act. 1877. Myanmar Code Vol XI.

The Sea Custom Act, 1877. Myanmar Code Vol III.

The Penal Code, 1860. Myanmar Code Vol VIII.

Chapter 21

The Intellectual Property Regime in Sri Lanka and its New Trends

Merengnege Antony Mervyn Fernando
PG. Dip. (Energy) IESL Part I and II, C. Eng., MIE (SL)
Senior Mechanical Engineer, National Engineering Research and Development Centre of Sri Lanka
Ekala, Ja Ela, Sri Lanka
E-mail: mamf@nerdc.lk

ABSTRACT

Despite the continued claim that Intellectual Property Rights facilitate research activities and encourage technology transfer, the impact of Intellectual Property Rights on socio-economic development process of developing countries has evidently reflected in many areas, including health, agriculture and education. The developing countries have faced the challenge of constraint optimisation on how to implement the World Trade Organization Trade Related Aspects Intellectual Property Rights Agreement in such a way to minimise the socio-economic costs and maximise the national benefits. The third world states are now facing increased pressure toward higher standards of Intellectual Property Rights protection (i.e. the so-called TRIPS-plus). The attempts of the developed countries to evolve the TRIPS-plus regime, which appears in the form of Free Trade Agreement, provide opportunities for those countries to negotiate rules and commitments that go beyond what was not possible on the multilateral level. By entering into an Free Trade Agreement with the developed countries, the developing countries see some advantages in tariff reductions of agricultural, clothing and other products, but at the same time it closes down the opportunity for the latter to put forward the issues of their concern through the World Trade Organization including the harmonisation of TRIPS and Convention on Biological Diversity, access to medicines, and protection of genetic resources, farmers' rights and traditional knowledge.

***Keywords**: Intellectual property rights, Sri Lanka, Genetic resources, Traditional knowledge.*

Introduction

Intellectual Property as a concept compasses a set of legally enforceable rights emanating from intellectual creations, i.e., creations of the human mind. The law relating to intellectual property is concerns such creations and their protection and management. Intellectual property is considered to be a form of property, with some unique features of its own. It shares several of the characteristics associated with real and personal property. For example, intellectual property is an asset and has a monetary value. It can, like any other form of property, be owned, transferred, sold or licensed. The proprietor of intellectual property has the right, subject to certain restrictions, to use and alienate his intellectual property and to restrain others from encroaching upon his rights. Intellectual property is a kind of intangible property as it may not be identified or defined by its own physical parameters. However, to be protected, it must be expressed in a discernible form or way. Intellectual property is traditionally divided into two sections, namely, industrial property and copyright. Industrial property embraces the rights relating to areas such as inventions, industrial designs, marks and protection against unfair competition. The rights relating to literary and artistic works fall under the category of copyright while the rights relating to performances of performing artists, phonograms and broadcasts are called related rights.

The introduction of the western concept of intellectual property to Sri Lanka occurred during the British colonial rule. The British Inventor's Ordinance of 1859 was made applicable to Sri Lanka and the first Sri Lankan patent was granted on January 12, 1861. This law was replaced by the Patents Ordinance of 1906, which was purely based on the English Patent Law. The English Law of Trademarks was introduced to Sri Lanka in 1888 under the Ordinance No. 14 of 1888 and the first trademark was registered on 1st January, 1889. The Merchandise Mark of Ordinance of 1889 (Chapter 151) made provisions relating to fraudulent marks on merchandise. The Trademarks Ordinance (Chapter 150) of 1925 which was enacted in line with the English Trademarks Ordinance of 1919 replaced the Ordinance No. 14 of 1888. The Design Ordinance (Chapter 153) which provided for the registration of designs was passed in 1904. The English Law of Copyright was applied per se in Sri Lanka including the English Copyright Act of 1911. A statute called the Copyright Ordinance was enacted in 1912 to supplement the said English Copyright Act of 1911. There were some other statutes in existence in Sri Lanka relating to intellectual property such as the Patents, Designs, Copyright and Trademarks (Emergency) Ordinance (Chapter 157) and the Patents, Industrial Designs and Trade marks (The Neuchatel Agreement) Ordinance (chapter 156)

The Present Regime

An effective system of Intellectual Property became a pressing need in the recent past, especially in the context of the liberalized economic policies of the country. Consequently, the present Intellectual Property regime including law and administration as provided in the Code of Intellectual Property Act No. 52 of 1979 (which is hereinafter referred to as "the Code") was established.

The code was enacted to revise, consolidate, amend and embody in the form of a Code, the law relating to copyright, industrial designs, patents, marks and unfair competition and to provide for better registration, control and administration thereof and for matters connected therewith or incidental thereto. It is divided into eight parts. They relate to administration, copyright, industrial designs, patents, marks, trade names and unfair competition, offences and penalties and miscellaneous (which includes application to and proceedings before the Director and the Court and the Minister's power to make regulations, the Intellectual Property Advisory Commission and the Intellectual Property Fund).

Five amendments have been so far introduced to the Code by the Code of Intellectual Property Act Nos. 30 of 1980, 2 of 1983, 17 of 1990, 13 of 1996 and 40 of 2000.

Administration System

The Code provides for some fundamental matters relating to the administration of intellectual property including the establishment of a new Intellectual Property Administrative System. An Intellectual Property Office named "the Registry of Patents and Trademarks" was for the first time in Sri Lanka, established in 1982. The Registry of Patents and Trademarks was renamed "the National Intellectual Property Office of Sri Lanka" (NIPOS) by the provisions of the Code of Intellectual Property (Amendment) Act No. 13 of 1997. This Office is headed by an Official called "the Director of Intellectual Property" whose functions relating to Industrial Designs, Patents and Marks are *quasi* judicial in nature and subject to appeal to Court. The Office of Director of Intellectual Property is therefore judicially autonomous.

Objects of Intellectual Property Protected Under the Code

The Code offers protection to the rights relating to inventions, trademarks and service marks, industrial designs and literary and artistic works. In addition, it protects trade names and provides for the protection against unfair completion.

Patents

Patents are granted for patentable inventions which may be, or may relate to, a product or a process in a field of technology. An invention is patentable if it is new, involves an inventive step and is industrially applicable. The following, notwithstanding that they are inventions within the legal definition, are not patentable; discoveries, scientific theories and mathematical methods; plant or animal varieties or essentially biological processes for the production of plants or animals, other than micro-biological processes and the products of such processes; schemes, rules or methods for doing business performing purely mental acts or playing games; methods for the treatment of the human or animal body by surgery or therapy, and diagnostic methods practiced on the human or animal body.

The right to a patent belongs to the inventor or inventors. An invention made by an employee or pursuant to a commission belongs to the employer or to the person who commissioned the work provided that the parties have not agreed otherwise. A patent is valid for 15 days after the date of grant. After two years of the grant, the patent must be renewed each year until the expiration of the said period of 15 years.

Industrial Designs

A design is protected if it is new and not scandalous or not country to morality or public order or is not likely to offend religious or racial susceptibilities. Any composition of lines or colours of any three-dimensional form, whether or not associated with lines or colours, that gives a special appearance to a product of industry or handicraft is deemed to be an industrial design. A new industrial design means an industrial design which had not been made available to the public anywhere and anytime whatsoever through description, use or any other manner before the date of an application for registration of such industrial design or before the priority date validly claimed in respect thereof the right to obtain the registration of an industrial design belongs to its creator or his successor in title. A design created by an employee or pursuant to a commission belongs, unless otherwise agreed by the parties, to the employer or to the person who commissioned the work as the case may be the registration of a design is valid for a period of five years and renewable for two more consecutive periods of five years.

Marks

A mark means a trademark or service mark. A trademark means any visible sign serving to distinguish the good of one enterprise from those of other enterprises whereas a service mark means any visible sign serving to distinguish the services of one enterprise from of other enterprises. The exclusives right to a mark under the Code is acquired by registration. The registration of a mark may be granted only to the person who has first fulfilled the conditions of a valid application or who can validly claim the earliest priority for his application. The grounds of non-registrability on objective grounds and third party rights which include descriptiveness, non-distinctiveness, intrinsic and extrinsic deceptiveness, violation of the provisions relating to unfair competition, indication of source or appellation of origin and contravention of morality or public order years from the date of registration. The registration may be renewed for consecutive periods of ten years.

The Code does not totally deny the protection of unregistered marks. The use of a mark, even though it is not registered in Sri Lanka may play an important role in safeguarding the interests of the owners of such mark. Section 142 protects the unregistered marks within the framework of unfair competition.

Trade Names

A trade name, which means the name or designation identifying the enterprise of a natural or a legal person, is protected in Sri Lanka whether or not it is registered.

Protection against Unfair Competition

The concept of unfair competition is a new addition to the legal system of Sri Lanka. The Code while offering the legal protection against any act of competition contrary to honest practices in industrial or commercial matters sets out a list of acts of unfair competitions, which is not exhaustive, including all acts of such a nature as to create confusion by any means whatsoever with the establishment, the goods, services or the industrial or commercial activities of a competitor. As Fernando J.

observed in the Sumeet Research and Holding Ltd., case it is sufficient if what was done was in fact unfair in relation to the real competitor, whoever he was.

In term of Section 142(3) of the Code, any person or association of producers, manufactures or traders aggrieved by any act of unfair competition may institute proceedings in court to prohibit the continuance of such act.

Copyright and Related Rights

The rights of the authors of original literary, artistic and scientific works are protected under copyright. The Sri Lankan Courts have held that the originality relates to the expression of thought and that the expression need not be original nor in a novel form. The only requirement is that the work must not be copied from another work and it must originate from the author. What is protected under copyright is not the idea but the manner in which the idea is expressed. The law does not confer protection to a header of facts as well. The Code enumerates a list of works protected as literary, artistic and scientific works including writing, audio-visual works, musical works, works of drawing and painting, photographic works and works of applied arts. A recent amendment introduced to the Code has included computer program as a literary work in the list of protected works. (The Code of Intellectual Property Act (amendment) No. 40 of 2000) the list is not exhaustive. The derivative works such as translations, adaptation or other transformation of protected works, collections of works which constitute intellectual creations and works derived from Sri Lanka folklore are also protected as original works.

A work is protected automatically by operation of law. Registration or any such other formal requirement does not exist in Sri Lanka. Copyright protection in Sri Lanka extends to a wider area covering:

(*a*) Works of authors who are nationals of, or have their habitual residence in Sri Lanka

(*b*) The works first published in Sri Lanka

(*c*) All works, by virtue of treaty entered into by Sri Lanka are to be protected as well as Sri Lanka's folklore.

The law also recognizes certain statutory limitations to copyright including "fair use" which covers the areas such as the use of copyright works for educational and research purposes and non commercial private use.

The concept of related rights embraces the rights of the performers, producers of sound recordings and broadcasting organizations. The Code makes provisions only in respect of the rights of the producers of sound recordings.

Protected Rights

The owner of a patent has, subject to certain statutory restrictions, the exclusive right to exploit the invention, to assign and transmit the patent and to conclude license contracts. Any of the acts referred to above cannot be performed without the consent of the owner of the patent. The registered owner of an industrial design has the exclusive right:

(*a*) To reproduce and embody such industrial design in making a product,
(*b*) To import, offer for sale, sell or use of a product embodying such industrial design,
(*c*) To stock a product embodying such industrial design;
(*d*) To assign and transmit the registration; and
(*e*) To conclude license contracts.

The registered owner of a mark has, subject to relevant statutory provisions, the exclusive right to use the mark, assign or transmit the registration of the mark and conclude license contract. The third parties are precluded from the acts such as any use of a mark or of a sign resembling it in such a way as to be likely to mislead the public. The Code also protects both economic rights and moral rights of the authors of literary, artistic and scientific works. The economic rights include the right to reproduce the work, to make a translation, adaptation or any form of transformation of the work and to communicate the work to the public whereas the moral rights encompass the right to authorship and the right against distortion. The economic and moral rights are generally protected during the life of the author and for 50 years after his death.

License Contracts

The registered owner of a patent or design can enter into a contract, which is called "license contract" by which he grants to another person or enterprise a license to exercise his rights to the patent or the design. The owner of a mark can grant by way of a license contract, a license to use the mark for all or part of the goods or services in respect of which the mark is registered. The owner of copyright has the right to grant licenses to exercise his economic rights to the work.

Renunciation and Nullity

The registered owner of a patent, mark or design may renounce or surrender the registration by declaration in writing signed by him or on his behalf and submitted to the Director who will, on receipt of the declaration, record it in the Register and publish the same in the Gazette. The Court has power to declare the registration of a patent, design or mark null and void on any of the grounds as declared by law.

Assignment and Transmission

The application for registration or the registration of a patent, mark or design may be assigned or transmitted. The owner of copyright can transfer his economic rights and not the moral rights.

Priority

An applicant for registration of a mark, industrial design or patent can claim priority of an application earlier made in a convention country under the Paris Convention for the Protection of Industrial Property.

Infringement and Enforcement

The Code recognizes a series of statutory offences involving the infringement of intellectual property rights. These offences are declared to be cognizable and bailable within the meaning of the Code of Criminal Procedure Act No. 19 of 1979. Infringement of Copyright, industrial designs, patents and marks, false representation regarding industrial designs, patents and marks and unlawful disclosure of information relating to patents are some of such offences. The infringer is liable to be punished with a fine or imprisonment or both.

Such infringement may also result in civil litigation whereby the infringer may be restrained by an injunction and ordered to pay damages to the affected party. Importation of goods in violation of the rights protected under the code is prohibited and governed by the provisions of the customs ordinance.

Some attempts have been recently made to streamline the civil litigation system in intellectual property by establishing a special court for intellectual property and simplifying the appellate process. Under the Code only the District Court of Colombo had the first instance civil jurisdiction over intellectual property. In terms of the High Court of the provinces (Special Provisions Act No. 10 of 1996, the High Court of the Western Province now exercises the first instance jurisdiction over intellectual property. The appellate process is simplified by permitting to challenge the orders or decisions of the High Court of the Western Province only in the Supreme Court.

International Obligations

Sri Lanka has been a member of several international conventions and treaties on Intellectual Property. They are the Paris Convention for the protection of industrial property (since 1952), The Madrid Agreement for the repression of false or deceptive indication of source on gods (since 1952), The Nairobi Treaty for the protection of Olympic Symbol (since 1984), the Patent Cooperation Treaty (since 1982), The Berne convention for the protection of literary and artistic works (since 1959), the Universal Copyright Convention (since 1983), The Convention establishing the (since 1979), the Agreement on TRIPS under WTO system (since 1995) and the Trademark Law Treaty (since 1996).

Towards The Future

Corrective Measures

Several provisions of the Code need reconsideration and corrective measures. For example, the law of copyright needs to be totally restructured as the existing provisions are outdated and inadequate in the present context. An industrial design is registered without substantive examination. This has facilitated fraudulent activities creating some unwarranted problems for the genuine creators and owners of industrial designs. It is thus necessary to consider where the law relating industrial designs needs improvements.

New Norms

The introduction of new norms to the Intellectual Property Law may occur mainly under the TRIPS Agreement.

The introduction of new norms to the Intellectual Property Law may occur mainly under the TRIPS Agreement.

Both industrialized and developing countries have entered the 21st century with somewhat uniform law of intellectual property as provided under the Agreement on Trade Related Aspects of Intellectual Property Rights (TRIPS) under Uruguay Rounds of Trade Negotiations. Being a party to the agreement on TRIPS Sri Lanka is required to change the law in line with its stipulations.

The legal norms covering the following areas have to be newly introduced to the Sri Lanka law under the TRIPS Agreement.

Copyright and Related Rights

The law relating to copyright protects the interest of creators of original literary and artistic works by giving them property rights over their creations whereas the law of related rights or neighbouring rights protects the legal rights of persons, natural or legal, who either contribute to making such creative works available to the public or produce works which show creativity or technical and organizational skill sufficient to justify recognition of a kind of property right. The term related rights embraces three kinds of rights–rights of the performing artistic, rights of the producers of sound recording and the rights of broadcasting organizations. As mentioned above the existing Sri Lankan law on copyright appears to be not comprehensive enough and therefore needs reconsideration and improvements covering *inter alia* the recently emerged areas in the light of information technology. New provisions have to be made in respect of the related rights.

Undisclosed Information

The protection of undisclosed information – technical information related to manufacture of goods or provision of services or business information which includes the internal information which an enterprise has developed so as to be used within the enterprise, is receiving world-wide recognition. Undisclosed information is protected if it enterprise, is receiving world –wide recognition. Undisclosed information is protected if it

(*a*) Is secret in the sense that it is not, as a body or in the precise configuration and assembly of its components, generally known, among or readily accessible to persons within the circles that normally deal with the kind of information in question;

(*b*) Has commercial value because it is secret; and

(*c*) Has been subject to reasonable steps under the circumstances, by the person lawfully in control of the information, to keep it secret.

Layout Designs of Integrated Circuits

An integrated circuit means a product, in its final form or an intermediate form, in which the elements, at least one of which is an active element, and some or all of the intended to perform and electronic function.

A layout design is synonymous with "topography" and means the three-dimensional disposition, however expressed, of the elements, at least one of which is a three dimensional disposition prepared for an integrated circuit intended for manufacture.

Geographical Indications

A geographical indications, like Ceylon Tea, is generally defined as an indication which identifies a good as originating in a territory, region or locality where a given geographical indications must be protected to prevent the use of such indication which would mislead the public as to the geographical origin of the goods. The existing Sri Lankan law does not protect such indications and therefore the introduction of new provisions is required.

Protection of New Plant Varieties (Breeder's Rights)

In terms of the provisions of the TRIPS Agreement, Sri Lanka is bound to protect new varieties of plants (breeder's rights) either by patents or by an effective *sui generis* system or any combinations thereof. The existing patent law of Sri Lanka prohibits the patenting of plants. The Acts of 1978 and 1991 of the International Conventions for the Protection of New varieties of plants (the UPOV Convention) establish minimum standards for the protection of plant breeder's rights. The protection is accorded to plant varieties that are distinct, novel, uniform and stable. The protection is accorded to plant varieties that are distinct, novel, uniform and stable. The standards of protection under UPOV principles are lenient than those of under the patent system. The plant breeder's right will also be a useful mechanism for the countries that are not willing to grant patents for plants and other living organisms. However, the 1991 Act of UPOV has introduced certain restrictive principles in several areas such as farmers' privileges.

Rental Rights

The concept of rental rights is another new area which attention. Under the TRIPS requirements, Sri Lanka is bound to introduce legal provisions giving the authors and their successors in title the right to authorize or prohibit the commercial rental to the public of original or copies of their copyright works at least in respect of computer programs and cinematographic works.

Other Areas

There are some other areas of the existing law which require changes under the TRIPS Agreement. For example, the valid period of a patent of a patent which is at present 15 years from the date of grant must be made minimum of 20 years from the date of application for the patent.

Enforcement

Enforcement of Intellectual Property Law is another area which needs constructive attention. The recent attempts to introduce somewhat expeditious enforcement mechanism as provided in the High court of the provinces (special provision act) have failed to achieve the desired objectives. The delay and cost of litigation is counterproductive in the process of socio-economic development. Thus, the existing enforcement mechanisms, both civil and criminal, should be re-organized. New provisions are required to promote counter measures by customs against exportation and importation of goods in violation of intellectual property rights.

Some Domestic Requirements

It has become necessary to introduce certain new norms mainly in order to rationalize the administration of Intellectual Property regime. For example, a better mechanism for the collective administration of copyright and reorganization of the profession of Intellectual Property Agents are crucial requirements in the today's context.

Other Future Events (Sri Lanka Inventors Commission)

Establishment of "National Centre for Inventions"

Aims

- ✰ To modernize Sri Lankan inventions to match the International market.
- ✰ To generate government revenue from foreign currencies to Sri Lankan inventions as international level products for the country's expected economic development.

Recognition of Creative Citizens

Aims

To get possible contributions from creative citizens for the potential economic growth of the country, lining them up through the "Rebuilding the Nation through invention project"'

"Rebuilding the Nation through Inventions";

Awareness and Encouragement Phase

To create Sri Lankan inventions and to recognize Sri Lankan hidden talent for the country's economic development.

Aims

- ✰ To invite gifted and creative citizens to and produce new inventions for the country's new development phase.
- ✰ To disseminate a message of the validity of the citizen's contribution through inventions for the country.
- ✰ To have active contributions of all intelligent people to the country's development phase.

International–Level Invention Exhibition in Sri Lanka

Professionally-arranged exhibition in Sri Lanka with innovative "Made in Sri Lanka" inventions for overseas communities.

Aims

To target the international business community to market "Made in Sri Lanka" product that are originated through the "Rebuilding the Nation through Inventions" phase and modernized by the "National Centre for Inventions".

New and Emerging Issues

The development of Intellectual Property Law is closely associated with the development of human creativity in general and the development of technology in particular. As a result, fast emergence of new issues in the law is a constant process and common occurrence. Some of the newly emerged or emerging issues which demand due attention may be briefly referred to as follows.

Information and Communication Technologies

New developments in information and communication technologies have given birth to several new issues in intellectual property law particularly in the fields of copyright and related rights. Those developments such as interactive digital networks and digital delivery, databases and computer-generated works have posed many challenges to the law and administration of Intellectual Property. Intellectual Property Rights are also of central importance in maintaining a stable and effective environment for the development of electronic commerce including the issues relating to domain names, trademarks, protection of well known marks and dispute resolution. The advent of new uses of works such as computer storage and retrieval, cable transmission, satellite transmission, reprography, home taping and rental have also resulted in the development of new norms. The WIPO Copyright Treaty and the WIPO Performances and Phonograms Treaty (both of 1996) have attempted to address some of these issues and to update the international protection of copyright and related rights in response to the challenges posed by digital technology, in particular by global networks such as internet. They contain, *inter alia,* the provisions which aim at preventing unauthorized access to and use of works such as books, music, songs and films on the internet or other digital technology, in particular by global networks such as internet. They contain, *inter alia,* the provisions which aim at preventing unauthorized access to and use of works such as books, music, songs and films on the internet or other digital networks. Sri Lanka is not a party to these two treaties but the embodied principles cannot be easily ignored. At the same time, many other issues such as protection of audio-visual performances, protection of the rights of broadcasting organizations and protection of data are emerging. It has become critically important today to study new issues and monitor the impact of digital technology, particularly of global networks such as Internet on law of copyright and related rights. In the modern environment of information and communication technologies, the methods of protection and management of rights have become dramatically complicated demanding new solutions.

Biotechnology and Biodiversity

Biotechnology concerns living organisms like plants and animals and non - living material like seeds and enzymes. Biotechnology and biotechnological inventions will have a significant impact on human life and society in the years to come because they relate mainly to the fields of medicine, food, energy and environment. The patentability of the inventions in the field of biotechnology is a controversial issue.

The protection of biodiversity has become another popular subject today. Protection of biodiversity and the control of so-called bio-piracy as well as the use of intellectual property in the protection of rights relating to bio-diversity are some of the issues that have received worldwide attention.

Traditional Knowledge

Traditional knowledge may encompass the knowledge held, developed and passed on from generation to generation by a particular nation, community of sometime a family. It covers a broader range of subjects, activities and practices such as art, literature, music, crafts and designs, environment, plants, animals, medicine and methods of treatment including rituals and useful properties of fauna and flora relating, *inter alia,* to medicine, food and food processing. In view of the involved subjects the traditional knowledge has a great scientific and commercial value and therefore is vulnerable to undue exploitation by outside forces mainly those who have financial resources and technological know-how. There were certain reported instances of attempted exportation of traditional knowledge related material from Sri Lanka. The protection of traditional knowledge does not fall within the ambit of the existing intellectual property regime.

The laws relating to various objects of intellectual property recognized under the existing law are not intended or devised to protect traditional knowledge. Traditional knowledge cannot be patented as it is already known and in the prior art. The law of industrial designs will not protect the designs in traditional art and crafts because they are not new as required under the law. The copyright law will not be applicable to the literary and artistic work in the traditional knowledge because they have been in existence for many years beyond the period of time during which the rights of the copyright owners are protected whereas the authors of such works are not known. So are the principles of law of undisclosed information. Traditional knowledge does not fit into the legal frame-work of undisclosed information which is purely based on private ownership and commercial interests. Certain holders of traditional knowledge, for example in Sri Lanka, do not use their knowledge for commercial purposes at all.

Consequently it has become important to examine carefully how the existing Intellectual Property System can protect the rights of the holders of should be mentioned that national laws of the protection of traditional knowledge would not generate desired results unless the international community is willing to respect traditional knowledge and internationally respected norms and standards are established.

Protection of Folklore

Protection of Folklore and management of rights relating to folklore have drawn the attention of many countries both developed and developing. The development of effective and workable standards for the protection of folklore and mechanisms for identification, management and enforcement of folklore rights are a subject of much importance. There can be links between the protection of folklore and of traditional knowledge and innovations and creations based on traditional knowledge. All these issues need careful and constructive examination. Section 12 of the Code makes provisions for the protection of works of Sri Lankan folklore but they are not very effective mainly due to lack of clarity.

Intellectual Property and Human Rights

Intellectual property right have become unprecedented relevant in the policy making process in various sectors such as trade, culture, investment, science and technology, environment and health. Intellectual Property can be a useful instrument in the development process of any given country but its role in the development process is not free from criticism and challenges. Against this backdrop, the issue relating to intellectual property and human rights is steadily emerging.

The human rights character of intellectual property is referred to in several international and regional instruments such as the Universal Declaration of Human Rights, the International Covenant on Economic, Social and Cultural Rights and International Covenant on Civil and Political Rights. For example, Article 27 of the Universal Declaration of Human Rights declares:

"Everyone has the right freely to participate in the cultural life of the community, to enjoy the arts and to share in scientific advancement and its benefits.

Everyone has the right to the protection of the moral and material interests resulting from any scientific, literary or artistic product of which he is the author."

The linkage as well as the possible conflicts between human rights and intellectual property rights is relevant in the areas such as right to culture, right to health, and right to equal opportunities and rights relating to traditional knowledge. It may be crucially important to achieve a balance between the intellectual property rights and the rights to development and to benefit sharing in the scientific progress and its application. A constructive debate on these aspects is timely.

Conclusion and Policy Recommendations

In light of the considerable and long-term efforts by developing countries to minimise the impact of the TRIPS Agreement, one might conclude that most developing countries oppose the high degree of IPR protection. That conclusion, however, is contradicted by widespread and enthusiastic support of many developing countries for entering into FTAs that demand for higher commitments on IPRs.

Given the fact that developing countries have often suffered from the weakening prices of raw materials, foods and semi-manufactured products, which are their main foreign exchange earners, any single developing country would have a strong incentive to sign an FTA with the US, as it is believed that such a treaty helps that country to secure access to the world's most lucrative market. However, by signing an FTA, the developing country agrees, in a binding treaty under international law, to respect any obligations contains in the agreement it has entered into. The treaty can be harmful to the country because it leads to a world in which TRIPS-plus obligations are imposed. In making decisions with respect to bilateral or regional deals, policy-makers will have to weigh the economic benefits of FTAs against the importance of protecting health and social interests of their population.

Although some sectors of the economy may gain benefits from the bilateral or regional trade deal, it should be recognised that the benefits are limited only for particular sectors and certain groups of interests. On the contrary, the long-term social and economic costs that result from IPRs commitment are significant, and should not be underestimated as they affect the majority of the population. Strengthening protection of IPRs, regardless of specific needs and social priorities of each country, may sharply reduce the developing countries' industrial and technological competitiveness and will give rise to stronger dependencies on the more powerful countries. In conclusion, we believe that increased national protection of IPRs should be made on the ground of its assistance for the promotion of national technological and economic development, rather than in exchange for the uncertain benefits under the FTA.

In addition to the options suggested earlier, the following recommendations are being made in order to reduce the impacts of the TRIPS-plus regime:

- ✰ Recognise the irreversibility nature of the FTA commitments and their long terms implications on the developing countries.
- ✰ Call for a moratorium on FTA negotiations until reliable impact assessment studies have been carried out.
- ✰ Any bilateral trade negotiations should be viewed as an opportunity to address a much broader range of concerns of the developing countries, including abuse of intellectual property rights, technology transfer and capacity building, protection of traditional knowledge, and controlling and regulating access to and use of genetic and biological resources.

Intellectual Property is today not something in sophistication but a part of life. It cannot be seen in isolation for its relationship to creativity, knowledge, information and economic growth. One of the major differences between developed and developing nations is the difference between success and failure in the creation and use of knowledge. Sri Lanka cannot run the risk of isolation, marginalization and being left behind the rest of the world in the knowledge-based environment where Intellectual Property regime is an indispensable constituent part. Against this backdrop, the intellectual property regime in Sri Lanka demands constructive and visionary attention of those concerned.

References

References from Journals

National Inventors Day, 14, March 2010 – New Programs Magazine

Sri Lanka Inventors Commission

References from other Literature

Dr. D.M. Karunaratne, Director – National Intellectual Property office of Sri Lanka,

"The Intellectual Property Regime in Sri Lanka And Its New Trends"

Reference from Websites

Jakkrit KUANPOTH, Sukhothai Thammathirat Open University UNCTAD/ICTSD/HKU/IDRC Regional Dialogue "Intellectual Property Rights (IPRs), Innovation and Sustainable Development" 8 – 10 November; Hong Kong, SAR, People's Republic of China

Chapter 22
The Status of Indonesian Intellectual Property Rights

Anny Sulaswatty
Head of Legal and Public Relations Affairs,
Indonesian Ministry of Research and Technology,
Indonesia
E-mail: anny-s@ristek.go.id

ABSTRACT

The problem concerning Intellectual Property Rights touches various aspects such as technological, industrial, and social and the cultural aspects. However the most important aspect if being connected with protection efforts for the work of the intellectual is the legal aspect. The legal aspects' focus was to overcome various problems emerging with regards to intellectual property rights. In this aspect the law should enable successful protection of intellectual work for the development and encouragement of the community's creativity. Many factors cause the development and advancement of national Intellectual Property Rights today is still far from satisfactory. In Indonesia the Intellectual Property Rights enforcement situation has improved to some extent. The Government's Intellectual Property Rights campaign including an Intellectual Property Rights action plan and the President's Executive Order establishing the National Committee on Intellectual Property Rights were identified as the main positive points. Indonesia has showed considerable political will to combat Intellectual Property Rights violations, e.g. through the establishment of a National Coordination Team on Intellectual Property Rights, a national task force reporting directly to the President. Moreover, interest in Intellectual Property Rights protection is increasing: national authorities appear committed to implementing and enforcing Intellectual Property Rights legislation, and senior staff is reported to be competent.

Keywords: *Intellectual property rights, Indonesia, Government, Legislation.*

Introduction

Intellectual Property Rights (IPR) are legal entitlements granted by governments within their respective sovereignties that provide patent, trademark, and copyright owners the exclusive right to exploit their intellectual property (IP) for a certain period. The basic rationale for IPR protection is to provide an incentive for innovation by granting IP owners an opportunity to recover their costs of research and development. Intellectual Property Law also protects the exclusive rights of owners to a variety of intangible assets such as music, literature, artistic works, discoveries and inventions. Patents, copyrights, trademarks, industrial designs, domain names and trade secrets are also covered under IP Law.

Generally, IPR could be divided into two categories that is: Copyrights and The Industrial Property Rights. Whereas the Right of the industrial Wealth or Industrial Property Right covered the Trademark, the design of the Industry, the design of the layout of the most United/integrated Circuits, the Secret of the Trade and the Variety of the Crop.

IPR became the important part in the development of the national and international economies. Various information on policy, regulation, the latest development of the practice of the application and the protection of intellectual property rights, became materials that really were needed by various social circles (community), like the academic, the professional group, the industry, and the government in the national and international scopes as shown in Figure 22.1.

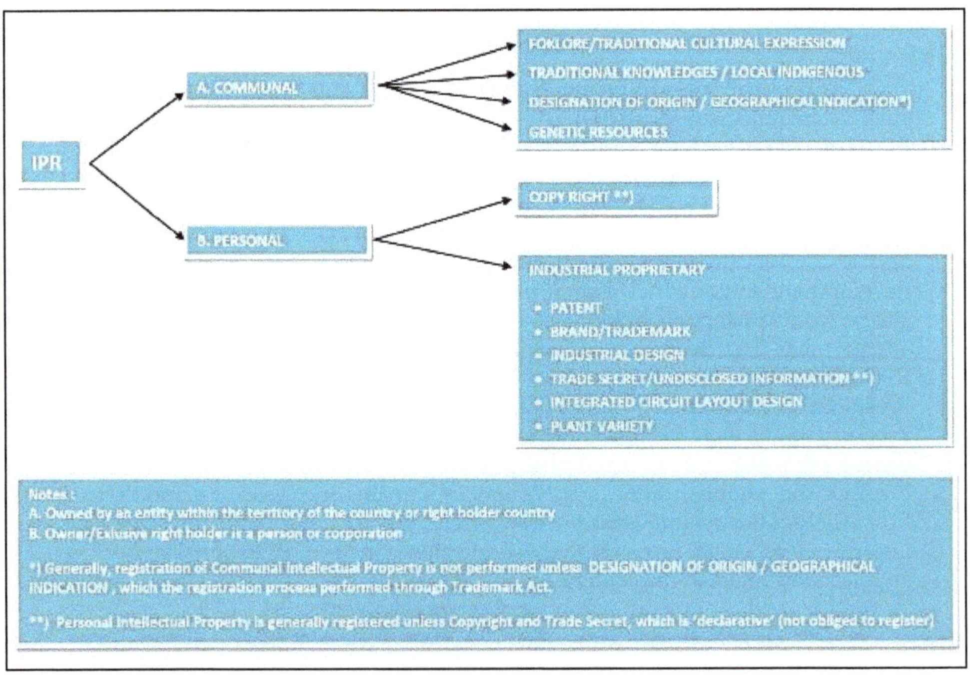

Figure 22.1: The Scopes of Intellectual Property Rights

Source: The Ministry of Research and Technology, Indonesia, 2010.

While many countries have some form of IPR protection, the degree to which IPR is protected varies greatly. For economists, IPR protection represents a trade off between the benefits of innovation and the costs of exclusivity. Property rights encourage the development of new technologies, products, music, and other creative output. Exclusivity, however, protects IP owners from competition and, in some cases, can grant monopoly power. For this reason, IPR protection is always limited in either the length or scope of protection. After expiration, for example, the knowledge embodied by a patent enters the public domain.

Information as the window and the main gate of knowledge had the dominant position in supporting the development, especially in the real sector of the community. The Directorate General of Intellectual Property Rights, Ministry of Law and Human Rights Republic of Indonesia as the executive and the task force of the government obligation to serve the requirement for the community would intellectual property rights information openly, was realized this official site.

The existence of IPR in relations between humankind and international was something that could not be ignored. IPR also was something that given and inherent in an industrial society or that was heading there. His existence always followed the dynamics of the development of the community personally. So also the matter with the community and the Indonesian nation that wanted did not want correlation and was directly involved with the IPR problem.

As the consequences of Indonesian participation as the WTO member; the World Trade Organization required Indonesia to adapt all of its legislation regulations in the intellectual property rights field with the standard of the TRIP's (Trade Related Aspects of Intellectual Property Rights) that was begun since 1997 and was renewed afterwards during 2000 and in 2001. This also resulting from being ratified by the international conventions in the intellectual property rights field but also adapted with the provisions that were required that is regulations about Copyright, Design of the Industry, Layout designs of the Integrated Circuit, Secret of the Trade, Patents and The Trade marks.

The problem concerning intellectual property rights touch various aspects such as technological, industrial, and social and the cultural aspects. However, the most important aspect if being connected with protection efforts for the work of the intellectual is the legal aspect. The legal aspects' focus was to overcome various problems emerging with regards to intellectual property rights. In this aspect the law should enable successful protection of intellectual work for the development and encouragement of the community's creativity.

The aspect of technology also was the factor that was very dominant in the development and the protection of intellectual property rights. The current development and dissemination of rapid information technologies is forever increasing thereby leaving behind countries that cannot catch up. The knowledge dissemination through IT ensures rapid information distribution worldwide. In the situation like this intellectual property rights (IPR) becomes increasingly important. IPR is the monopoly rights that are used to protect investments therefore ensuring that the inventor benefits from their creation.

The agency that had authority in carrying out intellectual property rights in Indonesia is the Directorate General of Intellectual Property Rights, which is under the Ministry of Law and Human Rights of Republic of Indonesia. The Directorate Information Technology under the Director General, IPR deals with IPR information. It is clear that the Indonesian government acknowledges the IPR as serious attention is given to them.

With the existence of an integral system of intellectual property rights information which is easy to be accessed by the community, there is evidence that the level of the registration requests for the Right of Indonesian Property in Indonesia increased. Law enforcement is an integral manner (where including inside intellectual property rights), and will decrease violation in the form of the hijacking of the work of the intellectual which is protected by regulations. Synergy between the IPR system and integral law enforcement of intellectual property rights information will in the long run bring the Indonesian nation to the more civilised life through honouring the work composed or created by other people.

Intellectual Property Rights Awareness and Application

The IPR component evaluates the protection of intellectual property. In addition to an opinion-based measure of the protection of IP, it assesses protection of two major forms of intellectual property rights (patents and copyrights) from *de jure* and *de facto* perspectives, respectively.

How is the level of the awareness of the community towards Intellectual Property Rights in Indonesia? According to the data, the following ratings were realised for IPR awareness in Indonesia: 7.5 per cent High; 13.76 Medium; and 78.74 Low. And the Indonesian Statistic of IPR is shown in the Figure 22.2.

Intellectual Property Rights Regulation

Attention and commitment of the Indonesian government to the acquisition, ownership, and protection of IPR has been defined in various legal and statutory regulations. The government is committed to advancing the patent and intellectual property rights contained in the Constitution (UUD 1945), especially the fourth amendment of Article 31 Paragraph 5 which states that *"The government advance the science and technology to uphold religious values and national unity to promote the civilization and prosperity of human"*.

In implementation, the government established IPR laws and regulations, such as:

(*a*) Act 19/2002 about Copyrights;

(*b*) Act 14/2001 about Patent;

(*c*) Act 15/2001 about Trademark;

(*d*) Act 31/2000 about Industrial Designs;

(*e*) Act 32/2000 about Layout Designs of the Integrated Circuits;

(*f*) Act 29/2000 about Plant Variety Protection;

(*g*) Act 30/2000 about Trade Secrets.

Figure 22.2: The Indonesian Statistic of Intellectual Property Rights (IPR)
Source: The Directorate General of IPR, 2010.

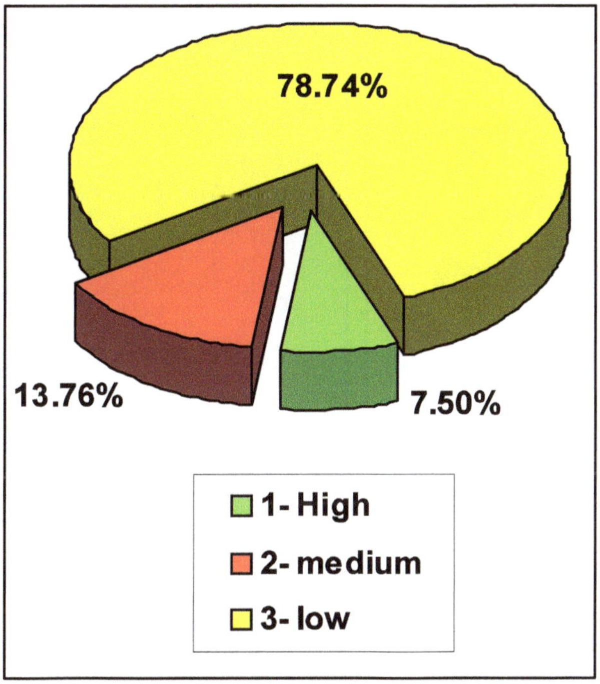

Indonesian IPR Awareness

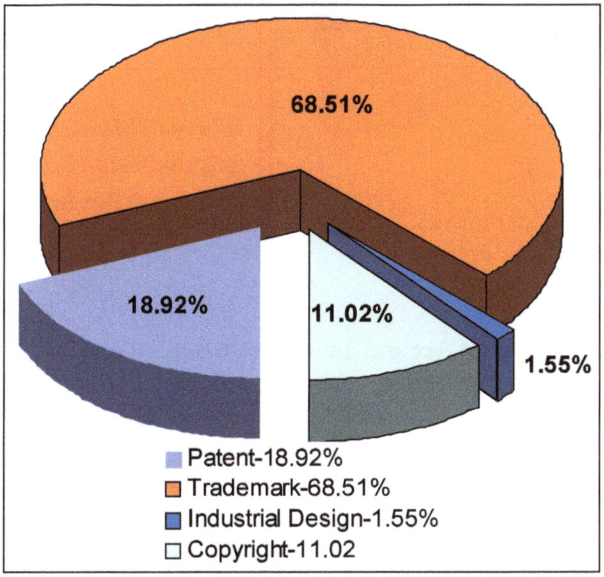

Indonesian IPR Application

Contd....

Figure 22.2–Contd...

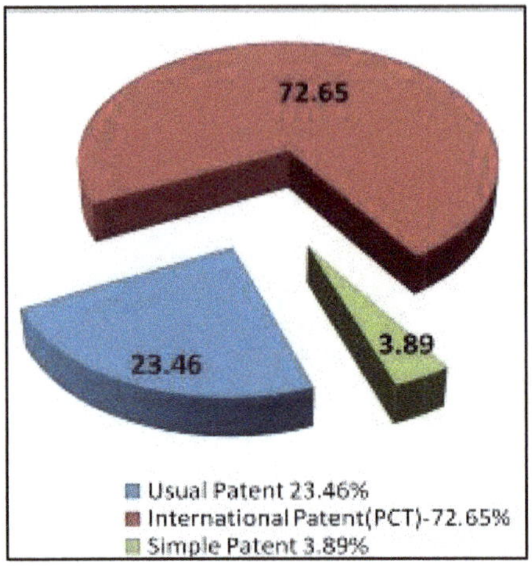

Patent Application

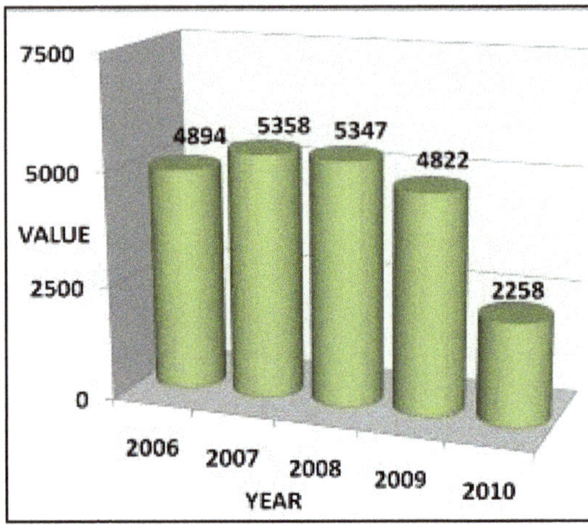

Patent Data Entry

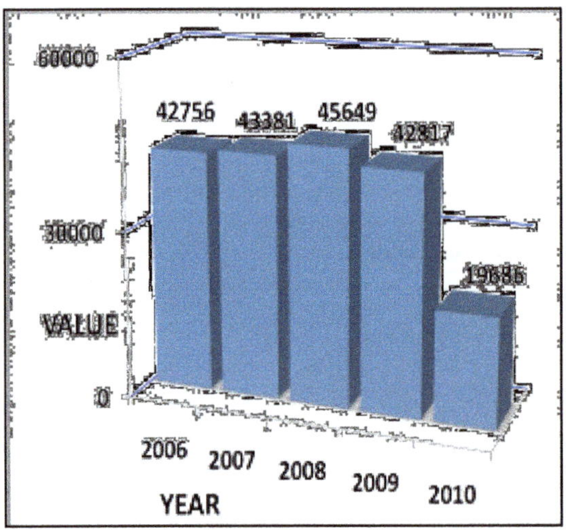

Brand Data Entry

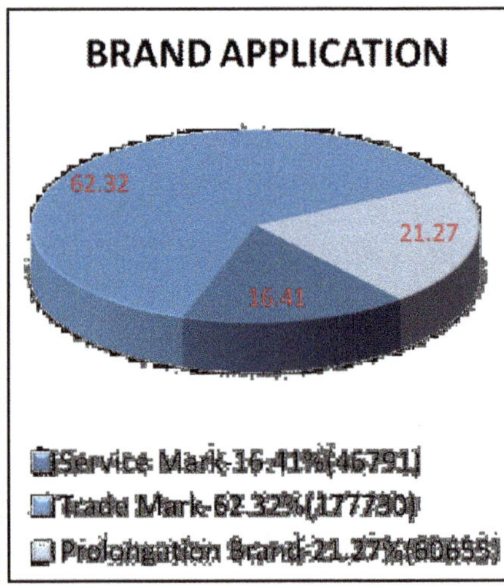

Brand Application

Contd....

In addition in the form of Act and government regulations, the national policy of intellectual property has also been prepared as one base in the development of intellectual property. This policy is specifically formulated by the National Response Team Violation of Intellectual Property Rights (Timnas-PPHKI) which was established by the Republic of Indonesia Presidential Decree Number 4 of 2006 which decided on March 26, 2006.

The rationale for the formulation of national policies in the field of intellectual property, are:

1. As principal guidance in establish the Act and government regulation about intellectual property rights;

Figure 22.2–Contd...

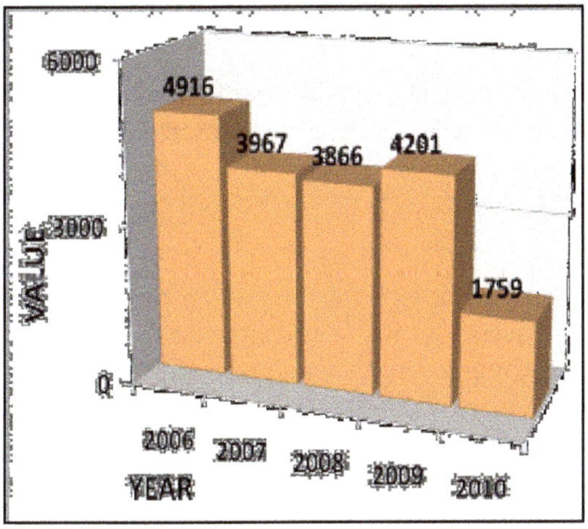

Design Data Entry

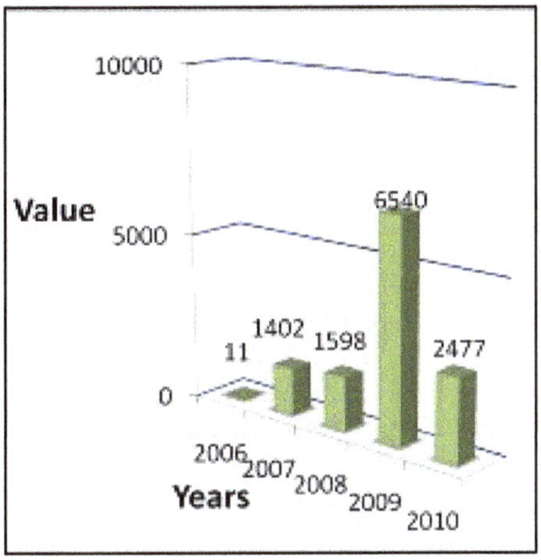

Copyright Data Entry

2. To facilitate in forming an environment that encourages the creation, protection, enforcement, management and exploitation of intellectual property as a driver of future industrial growth of the nation;
3. To promote and inculcate the culture of intellectual property in the business community and society. Also to create a community that respects the rights of IP owners and create a society that believes that intellectual property is a valuable economic asset.

Problems and Challenges of Facilitating the Acquisition Ownership of IPR

Many factors causes the development and advancement of national IPR today is still far from satisfactory. Based on the results of reading the literature and supported by the results of interviews with stakeholders, there are four principal issues in order to facilitate and simplify the process of obtaining patents and proprietary intellectual property rights of technology products and creative products. The four problems are as follows:

1. Regulation about patent license does not support;
2. Weak institutional and intermediary managers IPR;
3. Awareness of the importance of patent and other IPR; and
4. Weak link in the management of patents and IPR ownership.

The four problems mentioned above become increasingly complex as a result of the dynamics of the changes brought about by a national commitment to globalization and the commitment of Indonesian membership in the organization of world trade (World Trade Organization-WTO), especially regarding the setting up IPR

(Trade Related Aspects on Intellectual Property Rights - TRIPS). These arrangements provide a big impact on IPR protection and cause some challenges for Indonesia which include:

1. Adequate protection of IPR would encourage fair competition and otherwise ignore the protections in this area will lead to unfair competition and distrust of foreign investors;
2. Trade globalization demands transparency in the legal field and the creation of good legislation and can adequately protect IP owners;
3. Consistent attitude of the courts and authorities in the enforcement of such provisions becomes the object of scrutiny and international monitoring. Weakness in this sector will affect investment flows; it can even be used as the basis for the actions of replies from the countries who feel wronged. These actions may include sanctions in the economic field and trade (economic reprisal).

Therefore, Indonesia must harmonize all legislation in the field of IPR in accordance with Article 65 Agreement on Trade Related Aspects Intellectual Property Rights (TRIP's), in the following ways:

(*a*) Revision or change the laws that already exist in the field of IPR and to prepare new legislation for the field of IPR, such as Industrial Designs, Layout Designs of Integrated Circuits, and Trade Secrets;

(*b*) Restore and improve the quality of law enforcement in the field of IPR;

(*c*) Disseminating IPR;

(*d*) Improving the IPR administration and improve service to the community.

Facilitation of the Acquisition of Patents and IPR Ownership of Technology Products and Creative Products

Facilitation of the acquisition of patents and IPR ownership of technology products and creative products still poses a major problem and should receive attention including commitment of all stakeholders. This is because the consideration for providing protection to information and knowledge that is possessed, but also because of acquisition and ownership of IPR has a moral and economic values that can improve competitiveness and self-reliance including prosperity. This need has become increasingly important with globalization.

The presence of attention and commitment of the government through the establishment of the rule of law and legislation did not guarantee the smoothness of efforts to encourage community, researchers, engineers, business and science and technology community to obtain patents and the IPR. Therefore other important efforts are required to support the facilitation of the process of obtaining patents and IPR ownership as shown in Figure 22.3.

Several steps to improve the facilitation process of obtaining ownership of patent and IPR are as follows:

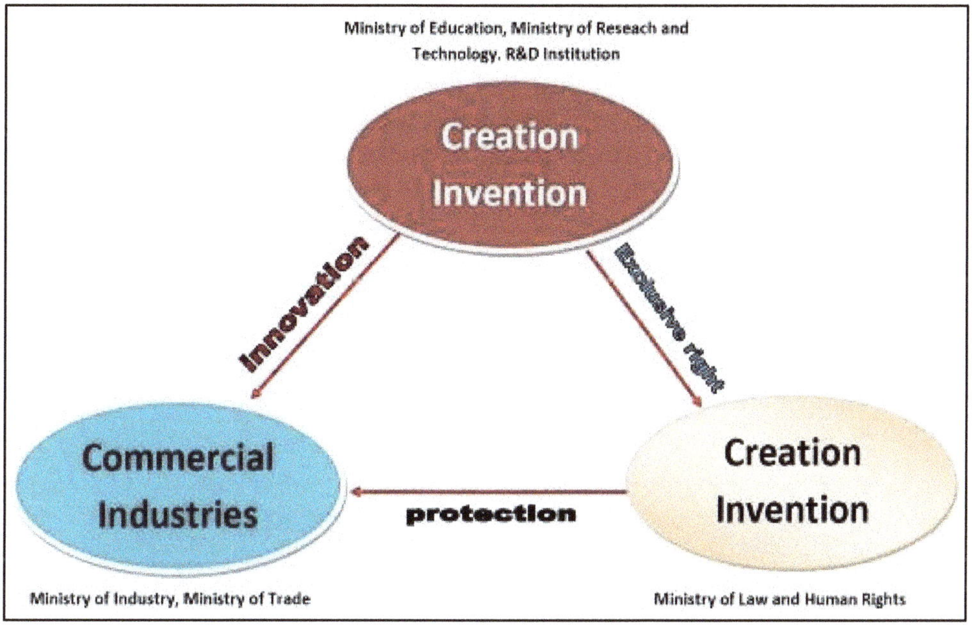

Figure 22.3: The Indonesian Intellectual Property Rights Related Institutions
Source: The Ministry of Research and Technology, Indonesia, 2010.

1. Encourage the issuance of license regulations in the field of patents;
2. Institutional strengthening IPR managers and intermediaries, including central IPR;
3. Improving the quality and quantity of resources;
4. Strengthening of IPR management network system.

Indonesian Country Profiles

Country profiles feature in the IPRI (International Property Rights Index) and its components' scores for the period 2007–2010 ranked in the index. They include commentary that identifies trends or changes in the overall component scores and the IPRI. Importantly, they also discuss any major advances or regressions in the sub-component scores. This allows readers not only to see larger trends but also to understand the driving factors behind these changes. Additionally, any particularly strong or weak features of the countries' property rights regimes, as measured in the index, are pointed out.

The country profiles are a complimentary tool to the IPRI rankings. They provide a snapshot of a country's performance in the IPRI over the last four years. As such, they should not be construed as a comprehensive overview of a country's property rights regime. However, the authors are hopeful that they will be used in combination with the rankings to bring attention to specific issues in property rights that need improvement. Alternatively, they can be used to commend any government policies contributing to positive developments in the protection and enforcement of property rights.

Where possible, the country profiles refer to policy developments in countries that could be associated with particular score changes. These are certainly not the only factors that could affect a property rights environment. They are provided here for informational purposes, and further review of such factors is encouraged.

Finally, the country profiles address any data inconsistencies that hinder the comparison of the IPRI and its sub-components' scores across years. Unfortunately, such inconsistencies persist because of the lack of some sub-component data.

The country profiles enhance the IPRI report and provide an additional tool to be used by policy makers, business leaders, think tanks, and other interested parties to better understand the situation with respect to property rights in their respective countries.

INDONESIA | **World Rank: 97** | **Regional Rank: 16**

The Indonesian IPRI score did not change in 2010, but the property rights environment in Indonesia underwent significant change.

The LP component increased dramatically for the third consecutive year. Every sub-component score increased. Judicial Independence rose an entire point.

The PPR and IPR scores offset any improvements in LP. While Protection of Physical Property Rights slightly improved, Access to Loans decreased an entire point, resulting in a net decrease for PPR.

The IPR score decreased because of negative developments in the Copyright Piracy score. Pirated material now is estimated to account for 85 per cent of the business software market and 95 per cent of records and music.

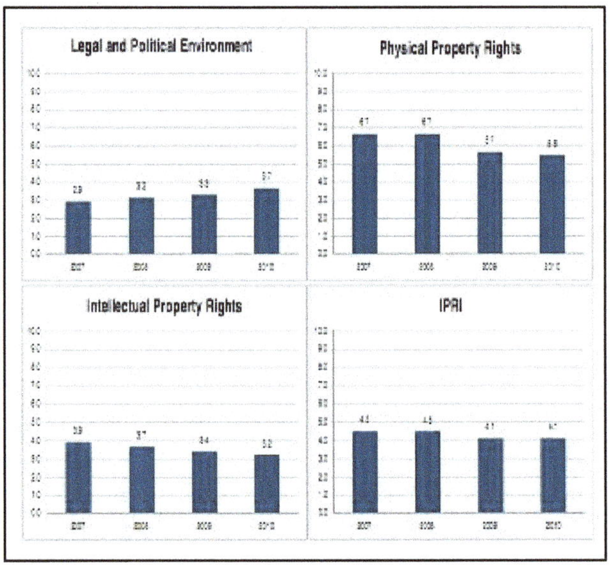

Source: International Property Rights Index, Report 2010

When referring to score changes, the names of the IPRI sub-components are capitalized and italicized. The reader is reminded that all IPRI scores are on a 0–10 scale.

Intellectual Property Rights Arbitration Agency

The Intellectual Property Rights World Day, which was commemorated by the World Intellectual Property Organization (WIPO) member countries in the previous

April, had the important meaning as Indonesia is one of the member countries of the organization.

At the time, several experts, the practitioner, the academic and 15 organizations/ the association of stakeholders in the field IPR initiated the idea of the formation of the Alternative Dispute Resolution (ADR) Agency.

This Agency was named the"Arbitration Agency and the Intellectual Property Rights Mediation (BAM HaKI)" which was supported by Minister of Law and Human Rights of Republic of Indonesia. This Agency will later be independent although being supported by the Law and Human Rights Minister, as well as the authority of IPR.

Despite only handling the civil cases, the independent's body can contribute the law enforcement IPR programme in the country. It was suggested that IPR at this time were assets for the transnational company which are deemed very important.

The presence of the Alternative Dispute Resolution (ADR) agency, at this time very much needed to complete/resolve the dispute outside the court's route. The arbitration Agency and the IPR mediation were also formed in Japan, China and Jordan.

Until now, the dispute of civil cases regarding the IPR was handled through the court's route. Prior to arbitration agency companies had a choice of resolving their own disputes. The dispute resolution outside the court's route, through arbitration and the mediation it was thought had several profits as the process of the case of being closed, was not open for the public like the session in the court. Moreover, the decision of arbitration was shaped like a final and binding, meaning that, the process was relative faster when compared with the process in the court because in the court had several stages of legal efforts when one of the sides was unsatisfied with the decision of the judge.

The first stage had equal efforts to the high court, the Supreme Court appeal to the Supreme Court and outside legal efforts could then be reconsidered. "In the arbitration agency the complaining parties did not have the equal term." the decision is final and binding, therefore relatively faster. Moreover, the handling of the case was carried out by the arbiters that indeed really the expert in the intellectual property rights field.

The originators or initiators of the formation of the arbitration agency and the mediation IPR hoped for the support from stakeholders in bringing about the independent's agent. The Agency such as ADR has been formed overseas, in fact WIPO has had one body of Mediation and Arbitration Centre to resolve the cases linked with IPR. The originators eager to bring and to realization about the same agency was provided a basis for by the message several legislation (UU) which are linked with IPR. It needs a fast judicature agency, but it does not mean that the available judicature agency in the district court was not fast, but still had a weakness for the case regarding IPR.

Till now, upholder of the law does not understand the IPR challenges however, with ADR/BAM HaKI, the arbiter would be able to deal with the IPR challenges. For example, in a case related to the biochemical patent, the case would be handled by experts in the biochemical field. The complainant's"sides" will feel calm and trust the system because the case is handled by a team that are experts in this field. The question now: Is whether this body will later become the choice for the industry to complete their dispute regarding IPR?. "Depended to stakeholders that personally". If they really wanted to obey the decision of the arbitration agency, the body will eventually become authoritative and trusted by industry. On the other hand, according to him, when the stakeholders did not heed the decision that was made by the agency, the presence of this agency will become nothing.

Selected 91 Patents to be Offered to the Businessman

The association of the manager ASPEKI of the intellectual's wealth carried out the selection towards 91 patents belonging to the PT tertiary institution to be offered to private companies.

The patent that will be offered to the private enterprise his status already granted. Meaning that, has had the certificate, was ready to be commercialised. "The patent will eventually be offered in one forum" for the "meeting between the inventor" of the "owner" of the "patent and the businessman around July." Aspeki would the focus to the patent in the pharmacy field.

ASPEKI had a membership of 121 centres and the intellectual property rights clinic of IPR University Research and Development Division, the tertiary institution and The Industrial Service Agency all over Indonesia. Those patents had the commercial element, but still not applied to the industry by the inventor. Although the forum for the meeting between the inventor and the industry' circle was carried out as a routine.

ASPEKI held a forum in March this year, but did not receive the anticipated response from the industry's circle. The failure of this forum, was because patents that were offered were owned by the perpetrators. Must be admitted, many patents have belonging to ASPEKI members were registered in Directorate General of IPR, but don't have industry application.

The lessons learnt by ASPEKI from all of that made it to carry out the mapping exercises for approximately any patent that was needed by the industry, for it to strike deals between inventors and the industry.

In Indonesia the IPR enforcement situation has improved to some extent. The Government's IPR campaign including an IPR action plan and the President's Executive Order establishing the National Committee on IPR were identified as the main positive points.

Indonesia has showed considerable political will to combat IPR violations, e.g. through the establishment of a National Coordination Team on IPR, a national task force reporting directly to the President. Moreover, interest in IPR protection is increasing: national authorities appear committed to implementing and enforcing IPR legislation, and senior staff is reported to be competent.

Despite these positive developments, serious problems regarding deficient enforcement of the domestic IPR regulations, high level of counterfeiting and piracy (for instance, it is estimated that 92 per cent of the video market is in the hands of pirates), uncertainty of the outcome of court proceedings and corruption have been identified as the main problems by the respondents to the Survey.

In addition, the implementation of certain TRIPs requirements does not yet appear to be complete, particularly in relation to the lack of legal protection of confidential information submitted by pharmaceutical companies to obtain marketing approval for drugs.

Appendix

The Indonesian Directorate General of IPR Activities in 2010 are:

A. Training, Seminar, Symposium and Workshop by WIPO

WIPO Training Course on the Examination Practice of Industrial Property, Japan, 14– 27 January, 2009

WIPO Training Course on the Industrial Property Administration, Japan, 28 January–10 February, 2009.

WIPO Training Course for Patent Examiners in the Optical Devices Technology, Japan, 25 February–3 March, 2009.

WIPO Training Course for Patent Examiners in Automotive Sector, Japan, 26 February–4 March, 2009.

1. WIPO Inter-Regional Forum for Heads of IP Offices and WIPO Roundtable for Heads of IP Offices on Current IP Issues, China, 30 March–1 April, 2009.
2. Long Term Trainees of WIPO Funds in Trust/Japan Program 2009, Tokyo, April– September, 2009.
3. WIPO Sub-Regional Seminar on Policy Options in the Application of Copyright Law, Singapore, 15–17 April 2009
4. Intermediate Seminar on Industrial property and Practical Training on Industrial Property, Austria, 10–26 June, 2009.
5. WIPO Asia Pacific Sub-Regional Seminar on Access to and Management of Copyrighted Works in the Digital Environment, Sri Lanka, 25–26 June, 2009.
6. WIPO Conference on IP and Public Policy Issues, Swiss, 13–14 July, 2009.
7. WIPO Asia Pacific Regional Symposium on Copyright Related Aspects of Information and Communication Technologies (ICT), Viet Nam, 29–31 July, 2009.
8. WIPO Observation Program on Madrid System in the Japan Patent Office (JPO), Japan, 4–6 August, 2009.
9. WIPO-SIPO Advanced Workshop on IP Management, Beijing, 25–28 August, 2009.

10. WIPO Regional Forum on the Role of National Design Councils for the Effective Promotion and Protection of Industrial Designs, Laos, 27–28 August, 2009.
11. Regional Workshop on Patent Information and Patent Information Conference: Building Success with IP, Singapore, 8–11 September, 2009.
12. Global Symposium of IP Authorities, Geneva, Swiss, 17–18 September, 2009.
13. WIPO Asia-Pacific Regional Seminar on Public-Private Partnerships (PPPs) in the field of IP for Developing Countries, Daejeon, Korea, 6–8 October, 2009.
14. Executive Workshop on the Application of Management Techniques in the Delivery of IP Services, Gatineau, Canada, 19– 23 October 2009
15. WIPO Regional Training Program on Technology Transfer and Successful Technology Licensing (STL) in the Energy Sector, Daejeon, Korea, 19–23 October, 2009.
16. WIPO-IP Australia Advanced Workshop on Examination of Traditional and Non-Traditional Trademarks, Canberra, Australia, 19–23 October, 2009.
17. Training Course on the Examination Practice of Industrial Property, Tokyo, Japan, 19–30 October, 2009.
18. WIPO Special Course on the Enforcement of Copyright and Related Rights, Tokyo, Japan, 19–30 October, 2009.
19. WIPO Conference on IP Strategy and Development, Manila, Phillipines, 27 October, 2009.
20. WIPO Asia-Pacific Regional Forum on Copyright and Related Rights, Singapore, 29–30 October, 2009.
21. WIPO Training Course on the Use of IT in Industrial Property Administration, Tokyo, Japan, 6–20 November, 2009.
22. WIPO Asia Pacific Regional Symposium on GI and Their Impact on Rural Economy, Chongqing, China, 30 November–1 December, 2009.
23. WIPO Automation Workshop for Asia and the Pacific Region on Smart IP Institutions, Singapore, 30 November–4 December, 2009.
24. WIPO Training Course on the Enforcement of IPR, Tokyo, Japan, 7–18 December, 2009.
25. WIPO Regional Seminar on the Protection of Biotechnological Inventions and the Budapest Treaty, Singapore, 8–9 December, 2009.
26. Study Visit to the Korea Copyright Commission, Seoul, 8–11 December, 2009.
27. WIPO Regional Seminar on Technology Transfer by Universities and Public Research Institutions through the Use of Patent System, Colombo, Sri Lanka, 9–11 December, 2009.
28. WIPO Training Course on the Examination Practice of Industrial Property (Basic Program), Tokyo, Japan, 13–26 January, 2010.

B. The Activity, Financed by The Directorate General of IPR (DJHKI) Budget

1. APEC IAP Peer Review, Singapore, 13 February, 2009.
2. The 28th APEC IPEG Meeting and Technology Transfer Seminar, Singapore, 23–26 February, 2009.
3. 13th Meeting of Standing Committee on the Law of Patents (SCP), Swiss, 23– 27 March, 2009.
4. Trade Investment and Facilitation Agreement-Trade Investment Council (TIFA-TIC), USA, 13–15 May, 2009.
5. 31st Meeting of AWGIPC, Thailand, 1-5 June, 2009.
6. Standing Committee on the Law of Trademarks, Industrial Designs and geographical Indications (SCT), Geneva, Swiss, 22– 26 June, 2009.
7. 4th Meeting on Intergovernmental Committee on Genetic Resources, Traditional Knowledge and Folklore (IGC-GRTKF), Swiss, 29 June– 3 July 2009.
8. APEC 29th Meeting of Intellectual Property Experts Group (IPEG), Singapore, 27–29 July, 2009.
9. The 47th Meeting Series of the Assemblies of Member States of the WIPO,Geneva, Swiss, 22 September–1 October, 2009.
10. Patent Cooperation Treaty (PCT) Union Assembly, Geneva, Swiss, 22 September–1 October, 2009.
11. National Team of PPHKI visit in the context of preparing a national database of SDGPTEBT, India, 26–30 October, 2009.
12. 32nd meeting oh AWGIPC, Siem Riep, Cambodia, 16–20 November, 2009.
13. Committee Meeting on Development and Intellectual Property, Geneva, Swiss, 16–20 November, 2009.
14. Standing Committee on the Law of Trademarks, Industrial Design and Geographical Indications, Geneva, Swiss, 23– 26 November, 2009.
15. Ministerial Meeting of WTO 7th, Geneva, Swiss, 29 November–2 December, 2009.

C. ASEAN Level IPR Activities

1. ASEAN-USPTO Workshop on Non-Traditional Trademarks and on the Implementation of the Madrid Protocol, Singapore, 10–11 February, 2009.
2. ASEAN-USPTO-USDOJ IP Crimes Enforcement Network (IPCEN) Conference, Thailand, 11–13 March, 2009.
3. ASEAN-USPTO Regional Patent Examination Training Program, Singapore, 31 March–1 April, 2009.
4. ASEAN-USPTO Colloquium for Public Prosecutors and the Judiciary on IPR Enforcement, Malaysia, 9–11 June, 2009.

5. ASEAN-USPTO Program on Technology Transfer, Thailand, 1–3 July, 2009.
6. ASEAN-APEC-USPTO Workshop on Effective Practices in the Border Enforcement of IPR, Honolulu, Hawaii, 20–23 July, 2009.
7. The ASEAN+3 Seminar on SMEs and Intellectual Property, China, 7–12 September, 2009.
8. ASEAN Training Seminar on PCT, Manila, Phillippines 12–16 October, 2009.
9. APEC-ASEAN-USPTO Seminar on Signal Piracy, Regulation and Enforcement in a Convergent World, Jakarta, Indonesia, 15–16 December, 2009.

D. Activities in Cooperation with European Patent Office (EPO)

1. EPA of EPO in 2009 WIPO/EPO/DPMA Inter regional Seminar on Industrial Property: Introduction to Patenting in Europe, Swiss and German, 10–26 June, 2009.
2. EPA of EPO in 2009 Patent Information (Beginners), Austria, 14-17 September, 2009.
3. Europe-Asia Patent and Patent Information Conference, Singapore, 10–. 11 September, 2009.

E. Activities in Cooperation with Asia Pacific Economic Cooperation (APEC)

1. APEC Workshop on Conducting Effective IPR Public Education and Awareness Campaigns for SMEs, Australia, 1–3 April, 2009.
2. APEC IPEG Trading Ideas Seminar, Singapore, 30– 31 July, 2009.
3. APEC Project for Training IPR Information Facilitators using e-learning content, IP Xpedite.

F. DJHKI Activities in Cooperation with WIPO

1. WIPO Regional Workshop on the Role of Information Technology in Effective Management of Intellectual Property Offices, Jakarta, 3– 5 February, 2009.
2. Advanced Training Course on Copyright and Related Rights in the Global Economy, Denpasar, Bali, 9–13 February, 2009.
3. Seminar and Exhibition on IPR, "Culture and High Technology and launching Indonesian Intellectual Property Academy (IIPA), FHUI campus UI Depok, 4 May, 2009.
4. National Seminar on Capacity Development for Small and Medium Enterprises (SMEs) through the System of Intellectual Property Rights (IPR) to Improve International Competitiveness, Jakarta, 25 June, 2009. Co-operation activities with RI Foreign Ministry.
5. Received a Visit from WIPO and JPO experts who conduct Quality Improvement Tokyo Mission Training Course, Tangerang, 22–23 June, 2009.

6. The Visiting of Mr Juneho Jang, Head Digitization Unit and JPO: Mr. Atsushi Koinuma, Assistant Director, Information System Division on behalf of WIPO Mission on Digitization, Tangerang, 7– 9 September, 2009 to observe the examination workflow application
7. WIPO Expert Mission on Copyright, Tangerang, 6–8 October, 2009.
8. PCT Training, Tangerang, 12–14 October, 2009). Cooperation with WIPO.
9. Branding Strategy Workshop, Jakarta, 26–30 October, 2009.

G. Activities within the framework of cooperation activities with Japan International Cooperation Agency (JICA), Japan Patent Office (JPO), The Association for Overseas Technical Scholarship (AOTS), Japan Institute of Invention and Innovation (JIII):

JICA Counterpart Training Course on "Intellectual Property Rights administration", Japan, 27 January–19 February, 2009; JICA Group Training Course on IPR, Japan, 6 May–11 July, 2009; JICA Training Course on Copyright System Development, Japan, 14 June–7 July, 2009; JPO/IPR Training Course for Lawyers, Japan, 29 June–17 July, 2009; Comparison Study of WIPO, Malaysia, 5–11 July, 2009; AOTS Training Course for Advanced IP Practitioners, Japan, 8–28 July, 2009; JIII/ AOTS JPO/IPR Training Course for Patent Experts, Japan, 26 August–15 September, 2009; JICA Group Training Course IP Administration for APEC Economies, Tokyo, Japan, 27 September–17 October, 2009. The JPO/IPR Training Program on Paten Examination Practice for ASEAN, Tokyo, Japan, 5 s.d.11 November, 2009. JPO/IPR Training Course for IP Trainer, Tokyo, Japan, 16 November–4 December, 2009. JICA Counterpart Training Course on Intellectual Property Administration, Tokyo, Japan, 27 January–18 February, 2010.

H. Activities Organized by Other International Organizations, whether they have done and will take Place

1. WTO Regional Workshop on TRIPs and Public Health, Singapore, 2–6 February, 2009.
2. International Law Enforcement Academy (ILEA), Thailand, 27 April– 1 May, 2009.
3. SIDA Advanced Training Course on Industrial Property in the Global Economy
4. 2009, Swedia, 1st 6– 28 April, 2009 and 2nd Regional Follow Up, Cairo, Mesir, 26– 30 October, 2009.
5. SIDA Genetic Resources and Intellectual Property Rights, Swedia, 4–22 May, 2009 and Follow-up Session in Pretoria, South Africa, 25–29 January, 2010. Korea-ASEAN Economy Cooperation Fund Training Course for the ASEAN.
6. Patent Examiners, Korea, 11–29 July, 2009.

7. Korea-Singapore Joint Training Program on IP Systems, Korea, 7–20 June, 2009.
8. FAO Seminar on quality linked to Geographical Origin, Bangkok, 8–11 June, 2009.
9. WHO E-Learning Course on TRIPs Flexibilities, Viet Nam, 24–26 June, 2009.
10. WHO Towards an IP Regime that Protects Public Health, India, 6–10 July, 2009.
11. The NUS Regional capacity Building in Intellectual Property Law Scholarship Program, Singapore, 27–31 July, 2009.
12. SIDA Copyright and Related Rights in the Global Economy, Swedia, Ist: 10– 28 August, 2009 and 2nd: Kuala Lumpur, Malaysia, 22–26 February, 2010.
13. CASBAA Convention and Associate Regulators Roundtable, Hong Kong, 3–5 November, 2009.
14. In-Went E-learning Course: Flexibilities of the TRIPs Agreement for Southeast Asia, Hanoi, Vietnam, 7–11 December, 2009.
15. Australian Leadership Awards (ALA) Fellowship Program, Canberra, Australia,March–May, 2010.
16. SIDA Training Program on Industrial Property in the Global Economy 2010, Stockholm, Sweden, 12 April–3 May, 2010.

I. The Activities Carried Out in Cooperation with Other International Organizations

1. DGIPR-WHO Workshop on the Examination of Pharmaceutical Patents: Developing a Public Health Perspective, Hotel athletes Century Park, Jakarta, 14–15 July, 2009.

www.ingramcontent.com/pod-product-compliance
Ingram Content Group UK Ltd.
Pitfield, Milton Keynes, MK11 3LW, UK
UKHW022212051026
15624UKWH00001B/122

9 789351 240846